T0215462

Communications in Computer and Information Science 1290

More information about this series at http://www.springer.com/series/7899

Snehanshu Saha · Nithin Nagaraj ·
Shikha Tripathi (Eds.)

Modeling, Machine Learning and Astronomy

First International Conference, MMLA 2019
Bangalore, India, November 22–23, 2019
Revised Selected Papers

 Springer

Editors
Snehanshu Saha ⓘ
CSIS and APPCAIR
Birla Institute of Technology and Science
Vasco Da Gama, Goa, India

Nithin Nagaraj
School of Humanities
National Institute of Advanced Studies
Bengaluru, Karnataka, India

Shikha Tripathi ⓘ
PES University
Bangalore, India

ISSN 1865-0929   ISSN 1865-0937   (electronic)
Communications in Computer and Information Science
ISBN 978-981-33-6462-2   ISBN 978-981-33-6463-9   (eBook)
https://doi.org/10.1007/978-981-33-6463-9

This Springer imprint is published by the registered company Springer Nature Singapore Pte Ltd.
The registered company address is: 152 Beach Road, #21-01/04 Gateway East, Singapore 189721, Singapore

Preface

We are immensely pleased to present the proceedings of the first international conference on Modeling, Machine Learning and Astronomy (MMLA), held during November 22–23, 2019 at PES University, Bangalore. The conference was organized by the Department of Computer Science and Engineering, PES University, in technical collaboration with the IEEE Computer Society Bangalore Chapter and the International Astrostatistics Association. The theory of machine learning, deep learning in particular, has been witnessing an explosion lately in deciphering "black-box approaches". Optimizing deep neural networks is largely thought to be an empirical process, requiring manual tuning of several parameters. Obtaining insights into these parameters has gained much attention lately. The conference focused on gaining theoretical insights into the computation and setting of these parameters and solicited original work reflecting the influence of the theoretical framework on experimental results on standard datasets and architectures. It is heartening to note that the conference was able to garner valuable talking points from optimization studies, another aspect of deep learning architectures and experiments. In this spirit, the organizers wished to bridge metaheuristic optimization methods with deep neural networks and solicited papers that focus on exploring alternatives to gradient descent/ascent-type methods. Papers with theoretical insights and proofs were particularly sought after, with or without limited experimental validation.

The conference hosted a stellar assembly of Keynote and Plenary speakers, renowned internationally for their contributions thematic to the conference. Pre-conference tutorials, which included a couple of hands-on sessions, attracted a lot of interest among the young audience.

Data is at the heart of this. Astronomy is a fascinating case study as it has embraced big data exemplified by many sky surveys. The variety and complexity of the data sets at different wavelengths, cadences etc. imply that modeling, computational intelligence methods and machine learning need to be exploited to understand astronomy. The importance of data-driven discovery in Astronomy has given birth to an exciting new field known as **astroinformatics**. This inter-disciplinary study brings together machine learning theorists, astronomers, mathematicians and computer scientists, underpinning the importance of machine learning algorithms and data-analytic techniques.

The Conference aimed to stake out unique ground as an amalgamation of the above diverse ideas and techniques while staying true to the baseline. The conference enabled discussion on new developments in modeling, machine learning, design of complex computer experiments and data-analytic techniques which can be used in areas beyond astronomical data analysis. Given the horizontal nature of MMLA, we hope we were able to disseminate methods that are area-agnostic but currently of interest to the broad community of science and engineering.

MMLA had three tracks: Modeling and Foundations, Machine Learning Applications and Astronomy and AstroInformatics. The three tracks attracted 63 research articles from India and abroad, out of which 16 were eventually accepted. We enforced a rigorous double-blind peer review system.

As volume editors, it was a privilege to be associated with the conference, which stood out because of its unique nature. We thank the Technical Program Committee for their dedicated efforts in reviewing the papers. We are thankful to the series editors of the Springer Book Series on *Communications in Computer and Information Science (CCIS)* for their support to bring out these proceedings of MMLA 2019. We thank all the authors of MMLA 2019.

November 2019

Shikha Tripathi
Nithin Nagaraj
Snehanshu Saha

Organization

Chief Patron

M. R. Doreswamy   PES University, India

Patrons

D. Jawahar   PES University, India
Ajoy Kumar   PES University, India

Steering Committee

K. N. Balasubrahmanya   PES University, India
   Murthy
V. Krishna Murthy   PES University, India
K. S. Sridhar   PES University, India
J. Suryaprasad   PES University, India
T. S. B. Sudarshan   PES University, India

Advisory Committee

Shikha Tripathi (Chair)   PES University, India
Somak Raychaudhury   Inter-University Centre for Astronomy & Astrophysics, India
B. S. Daya Sagar   Indian Statistical Institute, India
Bhabatosh Chanda   Indian Statistical Institute, India
Ajit Kembhavi   Inter-University Centre for Astronomy & Astrophysics, India
K. V. Subramaniam   PES University, India
Vishwas Lakkundi   IEEE Computer Society Bangalore Chapter, India
Saroj Meher   Indian Statistical Institute, India
Karthik Muralidharan   IEEE Computer Society Bangalore Chapter, India
Hima Patel   IEEE Computer Society Bangalore Chapter, India
James Peterson   Clemson University, USA
Vaskar Raychoudhury   Miami University, USA
Reetinder Singh Sidhu   PES University, India
Dinakar Sitaram   PES University, India
Rodolfo Valentim   Federal University of São Paulo, Brazil
Ben Wandelt   Sorbonne University, France

General Chairs

Snehanshu Saha   PES University, India
Jayant Murthy   Indian Institute of Astrophysics, India

Organizing Chair

S. S. Shylaja   PES University, India

Organizing Co-chairs

D. Uma   PES University, India
H. R. Mamatha   PES University, India
R. Jayashree   PES University, India

Publications Chairs

Shikha Tripathi   PES University, India
Nithin Nagaraj   National Institute of Advanced Studies, India
Snehanshu Saha   PES University, India

Finance & Publicity Chair

T. S. B. Sudarshan   PES University, India

Committee of Integrity

George Koshy (Chair)   PES University, India
Nithin Nagaraj   National Institute of Advanced Studies, India
Suresh Sundaram   Indian Institute of Science, India

Technical Program Committee

Saroj Meher (Chair)   Indian Statistical Institute, India
Nithin Nagaraj (Chair)   National Institute of Advanced Studies, India
Surbhi Agrawal   PES University, India
Stefano Andreon   Istituto Nazionale di Astrofisica, Italy
Suryoday Basak   University of Texas at Arlington, USA
Kakoli Bora   PES University, India
Francisco Förster Burón   Universidad de Chile, Chile
Guillermo Cabrera   Universidade Estadual de Campinas, Chile
Anirban Chakraborty   Indian Institute of Science Bangalore, India
Jessica Cisewski   Yale University, USA
Mousumi Das   Indian Institute of Astrophysics, India
Asif Ekbal   Indian Institute of Technology Patna, India
Eric Feigelson   Penn State University, USA

Contents

Astronomy and AstroInformatics

Modeling and Foundations

Optimizing Inter-nationality of Journals: A Classical Gradient Approach Revisited via Swarm Intelligence

Luckyson Khaidem[1], Rahul Yedida[2], and Abhijit J. Theophilus[3]([✉])

[1] State University of New York at Stony Brook, Stony Brook, USA
[2] North Carolina State University, Raleigh, USA
[3] Center for AstroInformatics, Modeling and Simulation (CAMS), Bengaluru, India
abhijit.theo@gmail.com

Abstract. With the growth of a vast number of new journals, the de facto definitions of Internationality has raised debate across researchers. A robust set of metrics, not prone to manipulation, is paramount for evaluating influence when journals claim "International" status. The ScientoBASE project defines internationality in terms of publication quality and spread of influence beyond geographical boundaries. This is acheived through quantified metrics, like the NLIQ, OCQ, SNIP and ICR, passed into the Cobb Douglas Production Function to estimate the range of influence a journal has over its audience. The global optima of this range is the maximum projected internationality score, or the internationality index of the journal. The optimization, however, being multivariate and constrained presents several challenges to classical techniques, such as curvature variation, premature convergence and parameter scaling. This study approaches these issues by optimizing through the Swarm Intelligence meta-heuristic. Particle Swarm Optimization makes no assumptions on the function being optimized and does away with the need to calculate a gradient. These advantages circumvent the aforementioned issues and highlight the need for traction on machine learning in optimization. The model presented here observes that each journal has an associated globally optimal internationality score that fluctuates proportionally to input metrics, thereby describing a robust confluence of key influence indicators that pave way for investigating alternative criteria for attributing credits to publications.

Keywords: Journal internationality score · Internationality production function · Particle Swarm Optimization (PSO) · Gradient descent · Machine learning

1 Introduction

There has been varying definitions of "Internationality" of peer-reviewed journals. Earlier research [1] on defining internationality claims that a journal qualifies to be called "international" if certain criterion are fulfilled. Concerning the

© Springer Nature Singapore Pte Ltd. 2020
S. Saha et al. (Eds.): MMLA 2019, CCIS 1290, pp. 3–14, 2020.
https://doi.org/10.1007/978-981-33-6463-9_1

vast geographical distribution of readers, journals are called "international" if their language of publication is English. According to the authors, any other native language used for publishing research articles may lead to lesser scientific impact and thus, low in international standards. Inclusion in the international databases, Impact factor and journal's editorial board members belonging to different nations are other parameters being used for measuring internationality as claimed by the authors.

Buchandiran [4] have reported an immense increase in articles and reviews published between 2004 and 2009. Evidently, with increase in articles, there is an emergence of vast number of new journals that claims to be international despite having low influence and bare minimum citations. Internationality tag attached to journals has often left many researchers in dilemma over how to carve out parameters that can measure a journal's influence in the right sense. Hence examining parameters from the ones that are popularly used in past years for modeling internationality of journals is an interesting and challenging problem to work on. Problem becomes even more challenging when the known parameters are found to be prone to opportunistic manipulations and gaming. Hence, there seems to be an explicit need to introduce new parameters that are robust, unbiased and can measure internationality of journal in an unambiguous way.

2   Related Work

Most of the work done on internationality deals with computing journal's influence by considering parameters like Impact factor, citations etc. Gunther K. H. Zupanc [15] claims that using Journal Impact Factor to compare influence is highly ill-suited since journals that belong to different domain may invite different citations and hence comparing two journals of dissimilar domains by considering just their Impact Factor is inappropriate. He claims that authors are tempted to publish their work in high-Impact Factor journals instead of journals that are best suited for their research work. In 2015, Neelam Jangid et al. [11,12] computed Journal Influence Score (JIS) by applying principal component analysis (PCA) and multiple linear regression (MLR) on scientometric indicators. The score is similar to the computing mechanism used in SCImago Journal and Country Ranking (SJR). Higher the score, the more likely that a journal is valued and accepted. JIS is a light-weight approach that uses MLR to compute score, whereas SJR uses a iterative Page Ranking Algorithm to compute score.

Source Normalized Impact per Paper (SNIP) is a very popular indicator introduced by Henk F. Moed [16]. It allows a fair comparison of journals that belong to dissimilar domains. The parameter uses Citation potential of all domains to normalize the Raw Impact per Paper of all journals. SNIP is defined as the ratio of the journal's citation count per paper and the citation potential in its subject field. It aims to allow direct comparison of sources in different subject fields. There is no single 'perfect' indicator of journal performance.

Different researchers across the world such as Anup Kumar Das, Sanjaya Mishra [14] contemplated on using article-level metrics (ALM) over Journal

Impact Factor (JIF) to assess the performance of individual scientists and their contributions. Seyyed Mehdi et al. [13] in their research made an attempt to study the scientific output of fifty countries in the past 12 years. A two-dimensional map is constructed, analyzed and studied in order to measure the 'quality' and 'quantity' of research output. Clusters are generated after analysis to represent country wise research.

3   Context of the Problem

Internationality has been extensively studied in ScientoBASE [5] where authors have defined internationality of journal in terms of quality of publications and in terms of spread of influence beyond defined geographical boundaries. A journal is more international when it encourages publishing scientific articles from authors/researchers which are from different demographic regions of the world. In order to measure journals on a scale of internationality, authors proposed novel metrics like NLIQ (Non-Local Influence Quotient), OCQ (other citation quotient), SNIP (Source Normalized Impact per Paper) and ICR (International Collaboration Ratio). These parameters, while computing internationality score, incorporates quality and influence in terms of publications and most importantly, invalidates all illegitimate attempts by journal to boost its influence/citations through unfair means. A brief note of these metrics is as follows. NLIQ is a metric that computes those citations that a journal receives from other journals of same or different domain in a specific time window. This metric captures non-localized impact that journals spread by publishing articles that maintain higher qualitative standards. OCQ is another parameter that find articles in a journal which are devoid of self citations. Self citation when practised by authors in large-scale, is often viewed as a strategy to unreasonably enhance their influence. ICR computes a ratio of authors collaborated from different nations while contributing articles in journal under evaluation. Essentially, ICR of a journal is high if a large portion of contributing authors in a journal are from different nations. SNIP allows comparison of journals of dissimilar domains. Few Journals that publish in niche areas tend to receive less number of citations as the publications are few in number. On the other hand, journal of an emerging domain receives higher citations mainly because of its reputation and popularity. Thus two journals from different domains cannot be compared solely on the basis of their raw impact factor because RIP computes influence on the basis of citations received by journals under comparison. To resolve this situation, SNIP computes citation potential of every domain, and divides journal's Raw Impact per Paper (RIP) with citation potential resulting in a score (SNIP), that allows fair comparison of journals from different domains.

Framework used in ScientoBASE [5] allows a fair computation of internationality score of journals evaluated across varying domains. A consolidated database consisting of parameters mentioned above, is built by using web scraping techniques and passed into Cobb Douglas Production Function. In Economics, the production function is used extensively to study the relationship of inputs with output and it is used for the first time in scientometrics to generate internationality index for journals. The function is given by

$$y = A \prod_{i=1}^{n} x_i{}^{\alpha_i} \qquad (1)$$

where y is the internationality index, x_i are the 4 derived journal parameters and α_i is elasticity. For close examination of its functional form, consider a case when 2 input parameters are used in the function (x_1, x_2),

$$y = A x_1{}^{\alpha} x_2{}^{\beta} \qquad (2)$$

Primarily, the function exhibits convex/concave properties for certain values α, β. These elasticity determines the response of output when the input levels are changed. There are three types of response depicted in output when inputs to the function are changed viz-a-viz: decreasing returns to scale (DRS, when $\alpha + \beta < 1$), constant returns to scale (CRS, when $\alpha + \beta = 1$) and increasing returns to scale (IRS, when $\alpha + \beta > 1$). Further, while experimenting with two inputs it was observed that, for certain values of α, β, function attains largest value of y, also called global maxima, that gave largest internationality score to journals. Appendix A explores the conditions for α, β, under which Cobb Douglas function is concave and achieves a global maxima, by examining the Hessian conditions for concavity of a function. In summary, Cobb Douglas is concave, when

$$\alpha \geq 0, \beta \geq 0, \alpha + \beta \leq 1 \qquad (3)$$

Modeling internationality of journals is a tricky problem. As explained above, the model of internationality [5,6,9] under the constant returns to scale constraint guarantees a theoretical optima. This is the optima we pursue as this indicator gives optimal internationality value for all journals which are found to have the reliable indicators/features used in the model. Please note that the model proposed is defined for three different types of constraints, CRS, DRS and IRS [5]. However, in our case, CRS constraint is chosen to solve the optimization problem. This is because the players (parameters of the model such as NLIQ) are in perfect competition toward the common goal of optimizing the production function i.e. internationality score of journals. It is pertinent to mention here that, the score function is split into two components with strong reasons to argue in favor of such approach. Each component comprising of scientometric indicators is in co-operative mode with the other so that the final internationality score is maximized. We know, the model is a concave function under the CRS constraint, thereby ensuring global optima. However, there are several issues that we intend to address here. To begin with, we need to address the fact that, even if theoretical optima exists, convergence of Newtonian methods (Gradient descent/ascent-fmincon in MATLAb)could be slow because of the multivariate nature of the function. Therefore, we need an approach which simulates gradients rather than computing those. Next, an optimal weighted combination of two components of internationality function (producing the score eventually) is often tedious and hard to find. Slight variations in weights could cause fluctuations in the internationality score. This is not desired. Therefore, we propose an approach where the final score is insensitive to the weights chosen in the sense that, a range of score (devoid of wide and wild variations) is obtained. This is the desired robust framework. Additionally, we want the solution to the optimization problem converge fast enough to the coveted global optima i.e maximum internationality score!

Perfect competition implies presence of a large number of parameters (scientometric indicators) driving a stable market equilibrium. Perfect competition essentially implies complete absence of inter-player competition (particularly of interest in our case since we don't wish parameters playing the domination game) because each is a small entity in view of the global internationality score, such that individual parameters (such as NLIQ, ICR) have little control or influence over the internationality function formation and the aggregate quantity. CRS in the usage of economics is integral to the presence of perfectly competitive markets by ensuring equi-proportionate returns to factor (parameter) inputs. Conversely, DRS implies that use of inputs generate less than proportionate increase in the output. Therefore to the extent that our indicators [1] optimize on profit or cost (Revenue which is our framework for internationality), these should resist expansion of production beyond the point where output (journal internationality score) grows less than proportionately to use of inputs. Since the concept of internationality score is borrowed from production economics we deem it necessary to interpret the objective in light of economics and therefore, we explore the bi-objective framework under CRS constraints rather than the DRS constraints. Therefore, the model under this particular CRS constraint provides an adequate motivation to explore the bi-objective optimization framework where the players, the components of the internationality function are in perfect competition with each other and ensures a Pareto front.

The above arguments augur well for exploiting PSO based optimization framework to ensure we reach the global internationality score of journals in robust and efficient manner. We proceed to explain the method in the next section which further illuminates the justification for using a metaheuristic such as PSO in place of classical Gradient ascent/descent methods. The paper is organized as follows. Section 4 describes the working of the traditional Particle Swarm Algorithm (PSO). Section 5 elaborates the algorithmic steps of PSO. Section 6 highlights the similarities of Gradient Descent and PSO. Section 7 examines the results and concludes the paper.

4   PSO: A Swarm Method to Simulate Gradients

The original authorship of PSO has been attributed to *Kennedy, Eberhart and Shi* [2,3]. PSO simulates the collective behavior of organisms in social context such as a bird flock or a fish school. The algorithm makes use of randomly

---

[1] Please note parameters, indicators, features and players have been used interchangeably in the manuscript.

initialized particles in an n-dimensional search space that traverse the fitness landscape to improve a defined measure of quality. This measure of quality is an objective function that needs to be either maximized or minimized. The particles in an swarm iteratively improve by learning from other particles in the swarm as well as from its own past experiences. A significant characteristic of PSO is that it does not make any assumption of the function being optimized and does not require any gradient calculation as in classical optimization techniques. The advantage is that it does not bear the overhead of calculating calculating gradient of complex objective functions in very high dimensional space. Also, it can optimize non-differentiable functions which would not be possible with gradient based classical optimization methods. It is worth noting that in classical optimization algorithms such as Gradient Descent, first order derivatives of the objective function is computed because it acts as a **guiding mechanism** for the solution to converge towards an optimum. This guiding mechanism is inherently captured in the way a swarm converges towards an optimum through social as well as experiential learning of every individual particle in the swarm. The paper attempts to provide theoretical insights to elucidate this subtle behavior.

5   Algorithm Description

Let $f : \mathbb{R}^n \to \mathbb{R}$ be an objective function. We need to find $x \in \mathbb{R}^n$ such that the function $f(x)$ is minimum. PSO first initializes a swarm of size s with random particle positions $X = \{x_1, x_2, x_3, ..., x_s\}$ where $x_i \in \mathbb{R}^n$ and $x_i \sim U(lb, ub)^n$. Here ub and lb are upper and lower bounds of the search space. Each particle is randomly initialized with velocity values $V = \{v_1, v_2, v_3, ..., v_s\}$ where $v_i \in \mathbb{R}^n$ and $v_i \sim U(v_{min}, v_{max})^n$. Here, $v_{min} = -|ub - lb|$ and $v_{max} = |ub - lb|$. v_{max} and v_{min} ensures that the velocity of particles are constrained within a specified range to ensure that the values do not explode and prevent or delay convergence. The algorithm maintains a set $L = \{l_1, l_2, ..., l_n\}$ where $l_i \in \mathbb{R}^n$ and $gbest \in \mathbb{R}^n$. l_i is current best observed position by particle x_i and $gbest$ is the current best observed position by the entire swarm. Given two particle positions x and y, x is said to be better than y if $f(x) < f(y)$ for a minimization problem. This means that at any iteration $f(gbest) \leq f(l_i) \; \forall \; i = 1, 2, 3..., s$. The particle positions at the tth iteration are updated in the following way:

$$v_{i(t)} = \omega . v_{i(t-1)} + r_1.c_1.(l_i - x_{i(t-1)}) \atop + r_2.c_2.(gbest - x_{i(t-1)})) \qquad (4)$$

$$x_{i(t)} = x_{i(t-1)} + v_{i(t)} \qquad (5)$$

ω is the inertia weight and controls the influence of the velocity from the previous iteration to the current velocity value. $c1$ is the cognitive factor which factors in the learning of individual particles from its past experiences and $c2$ is the social factor that models the tendency of each particle to move towards to the current globally best particle. The new particle positions are evaluated using the objective function and the set L and global best position *gbest* are updated as better solutions are found. This process is repeated until a termination criteria is met.

6   Gradient Simulation

To illustrate the simulation of gradients in Particle Swarm Optimization, it is imperative to bring out the similarities between PSO and classical optimization algorithms such as gradient descent. In the following subsection, the intuition behind gradient descent algorithm will be discussed.

6.1   Gradient Descent

Gradient descent minimizes a function $f : \mathbb{R}^n \to \mathbb{R}$ by first initializing a random solution $\theta \in \mathbb{R}^n$ and iteratively converges θ towards an optimum in the following way,

$$\theta_{(t)} = \theta_{(t-1)} - \alpha \cdot \frac{\partial f}{\partial \theta_{(t-1)}} \qquad (6)$$

where α is the learning rate which controls the extent to which the solution moves towards an optimum. Let us recall that the gradient of a function is $\frac{\partial f}{\partial \theta} = \lim_{\Delta\theta \to 0} \frac{f(\theta + \Delta\theta) - f(\theta)}{\Delta\theta}$. The gradient of a function measures the rate at which the function f changes when θ is increased by an infinitesimal value. And hence Eq. (6) implies that the solution moves towards the direction of decreasing gradient. The intuition here is that if $\frac{\partial f}{\partial \theta_{(t)}} < 0$ then, a minimum exists at $\theta > \theta_{(t)}$ and if $\frac{\partial f}{\partial \theta_{(t)}} > 0$ then, a minimum exists at $\theta < \theta_{(t)}$. Therefore, the gradient of a function at a particular point guides the solution towards a minimum in the search space. An interesting implication is that the gradient of a function at a point contains prior information on where an immediate optimum exists. Provided that α is small enough, the algorithm ensures that $f(\theta_{(t)}) < f(\theta_{(t-1)})$.

6.2   PSO and Gradient Descent: Equivalence

As mentioned before, PSO does not assume anything about the function being optimized and does not require any gradient calculation. And hence, PSO does not make use of the advantage of any guiding mechanism that the gradient of a function provides. It does not rely on any prior information where an optimum might exists to maximize or minimize a function. To overcome this, the algorithm focuses on collective behaviors that result from the interactions of particles or candidate solutions with each other and also through the experience of each individual particle. The initially randomized candidate solutions in PSO explore the search space and estimate initial guesses on where an optimum might exists. Then each individual particle factors in its own individual learning and also the learning from the best performing solution or particle to update its position in the next iteration. This particular characteristic of **Swarm intelligence** implies that each individual particle is guided by other members in the swarm. Over multiple iteration, the particles in the swarm converge towards a global optimum.

We know that at any iteration t, $f(gbest), f(l_i) \leq f(x_i) \ \forall \ i = 1, 2, 3, ..., s$. An interesting conclusion that can be drawn from this fact is that in Eq. (6), the terms $(gbest - x_i)$ and $(l_i - x_i)$ indicate directions in which a function is decreasing; meaning that the swarm iteratively moves towards an optimum as l_i and *gbest* are updated when better solutions are discovered. This particular characteristic of PSO is analogous to the way a candidate solution is improved in gradient descent by moving the solution in the direction of decreasing gradient. Over many iterations, the direction of each particle velocity converges with the direction in which the gradient decreases. With this, we can draw the conclusion that the velocity computed in PSO simulates the gradient of a function.

This equivalence has a fundamental connotation in the context of the Scientometric problem we're trying to solve. The internationality function of journals is written as a production function as mentioned above. Traditionally, Matlab or equivalent libraries (namely fmincon, a very popular routine) are used to compute the optima. This is done under the assumption that the function to be computed doesn't suffer from curvature violations and therefore the theoretical optima is not different from the computed ones. However, if the function continues to have additional parameters built in other than the four we have used here, the curvature violation will be imminent. This is expected as new features or parameters are defined by rating agencies and the metrics need to be updated accordingly. For a typical n parameter internationality function, where $n \geq 4$, curvature violation will ensure that Newtonian methods such as fmincon will not converge to the promised global optima even though theoretically there exists one under the problem and constraint settings of the optimization framework mentioned in the introduction section. Therefore, meta-heuristic methods such as PSO need to be exploited to compute global optima efficiently. Additionally, it is important to show that there is some equivalence between the classical and meta-heuristic approaches to dispel any doubt regarding guesswork!

7   Results and Conclusion

Conceptualizing a framework for journal internationality is necessary for quantification. Otherwise, the term is used loosely enough across communities to serve any real purpose. The work originated by Buela et al. is extended in this manuscript, where we intend to explore internationality of journals as a new concept that can be used to rank and classify journals based on levels on internationality. Once the ranking framework is built, the output will then be compared with standard journal rankings found in Scopus. The motivation of this exercise is to pave way for investigating the convergence between existing ranking system and our approach. Without valid and validated quality parameters, internationality of journals holds little merit. This hypothesis led us to elaborate investigation detailed in the manuscript. Indeed, without quality measures and

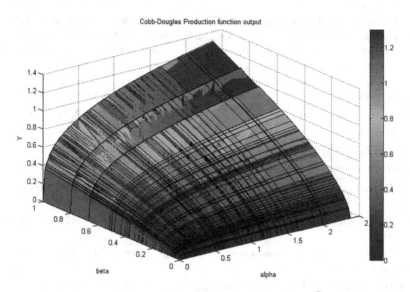

Fig. 1. Nonlinear profile of internationality function of journals: global maxima is the maximum internationality score.

quantifiable excellence criteria, internationality is merely a term not worthy of serious discourse. Much of the UGC guidelines focusing on "international" publications is therefore misplaced. This is the reason internationality needs to be studied in great detail so that a reasonable policy evolves regarding attributing credits for publishing in "international" journals.

Therefore, a model with valid theoretical background is presented where we observe that, a journal has a globally optimal internationality score and it may fluctuate in a manner proportionate to fluctuations in input parameters. This is guaranteed by the functional form of the model and a powerful tool in machine learning, known as convex optimization. The above figure captures such fluctuations and the global maximum internationality score is evident. However, computing the score presents a different set of challenges owing to multiple issues such as curvature violation, being stuck in local saddle points, scaling of parameters etc. Such problems are difficult to circumvent using classical, analytical techniques. Hence, the role of machine learning, as demonstrated in the manuscript, becomes paramount. We presented one such method, Particle swarm optimization (see Table 1) and exhibited the efficacy of the method in dealing with the difficulties of a complex model. The study of machine learning in Scientometrics, we hope, gains traction in the coming decade as it has become a necessary tool to analyze complex models from non-empirical standpoint. Classification of journals based on levels of internationality is another important task of supervised machine learning that should pose challenges. However, we maintain that, this is an important task to accomplish in future as it depends on the quality of machine learning algorithms and reliable data.

Table 1. Journals and respective metrics: the indicators, OCQ, NLIQ, SNIP, ICR and Internationality score computed by PSO: Our method handles multivariate nature of data and model complexity quite efficiently. The scores are between 0 and 1, with 0 and 1 being the lowest and highest bounds of internationality of journals.

Journal name	NLIQ	ICR	OCQ	SNIP	Int. Score
ACM Computing Surveys	0.98	0.11	0.98	1.71	0.83
Journal of the ACM	0.96	0.13	0.97	0.78	0.77
Journal of Data and Information Quality	0.83	0.11	0.89	0.17	0.58
Journal of Experimental Algorithmics	0.90	0.16	0.83	0.20	0.65
ACM Journal on Emerging Technologies in Computing Systems	0.80	0.11	0.77	0.19	0.57
Journal on Computing and Cultural Heritage	0.78	0.15	0.80	0.22	0.58
ACM Transactions on Autonomous and Adaptive Systems	0.93	0.18	0.82	0.37	0.71
ACM Transactions on Accessible Computing	0.85	0.11	0.87	0.41	0.65
ACM Transactions on Architecture and Code Optimization	0.90	0.15	0.87	0.56	0.72
ACM Transactions on Algorithms	0.95	0.22	0.82	0.50	0.76
ACM Transactions on Asian and Low-Resource Language Information Processing	1.00	0.13	0.83	0.15	0.66
ACM Transactions on Applied Perception	0.90	0.15	0.88	0.24	0.66
ACM Transactions on the Web	0.97	0.16	0.89	0.50	0.75

A   Appendix A: Necessary and Sufficient Condition for Concavity

Theorem 1. *For any function $f \in C^2; x \in R; f : R^2 \to R$ is concave iff the Hessian Matrix, $H \equiv D^2 f(x)$ is negetive semi-definite $\forall x \in U$.*

Proof. Here we are establishing that Cobb-Douglas function is concave under decreasing and constant returns to scale (i.e when $\alpha + \beta < 1$, $\alpha + \beta = 1$) and attains a global maxima when these elasticity conditions are met. We compute the Hessian Matrix of the function and show that the matrix is negative semi-definite when $\alpha + \beta < 1$ and $\alpha + \beta = 1$. This proves that Cobb Douglas is concave only when the above conditions on elasticity are met.

The function is given by $f(x_1, x_2) = k x_1^\alpha x_2^\beta$ with $k, \alpha, \beta > 0$ for the region $x_1 > 0$ and $x_2 > 0$. The Hessian Matrix (H) of the function is given by

$$H = \begin{bmatrix} \alpha(\alpha - 1)k x_1^{\alpha-2} x_2^\beta & \alpha\beta k x_1^{\alpha-1} x_2^{\beta-1} \\ \alpha\beta k x_1^{\alpha-1} x_2^{\beta-1} & \beta(\beta - 1)k x_1^\alpha x_2^{\beta-2} \end{bmatrix} \tag{7}$$

First order principal minors of H are:

$$M_1 = \alpha(\alpha - 1)k{x_1}^{\alpha-2}{x_2}^{\beta}; \quad M_1' = \beta(\beta - 1)k{x_1}^{\alpha}{x_2}^{\beta-2} \tag{8}$$

Second order principal minor is:

$$M_2 = k\alpha\beta x_1^{2\alpha-2}x_2^{2\beta-2}[1 - (\alpha + \beta)] \tag{9}$$

To prove that Hessian matrix H is negative semi-definite, we need to show $M_1 \leq 0$, $M_1' \leq 0$ *and* $M_2 \geq 0$. We know, for decreasing and constant returns to scale: $\alpha + \beta \leq 1$, therefore

$$\alpha \leq 1, \beta < 1 \Rightarrow (\alpha - 1) \leq 0$$
$$\Rightarrow M_1 \leq 0 \tag{10}$$

And also,

$$(1 - (\alpha + \beta)) \geq 0 \Rightarrow M_2 \geq 0 \tag{11}$$

Both conditions for concave function are satisfied by decreasing and constant returns to scale. Therefore, Cobb Douglas is concave, when

$$\alpha \geq 0, \beta \geq 0, \alpha + \beta \leq 1 \tag{12}$$

References

1. Buela-Casal, G., Perkakis, P., Taylor, M., Checha, P.: Measuring international-ity: reflections and perspectives on academic journals. Scientometrics **67**(1), 45–65 (2016)
2. Kennedy, J., Eberhart, R.: Particle swarm optimization. In: Proceedings of IEEE International Conference on Neural Networks, pp. 1942–1948 (1995)
3. Shi Y., Eberhart R.: A modified particle swarm optimizer. In: Proceedings of IEEE International Conference on Evolutionary Computation, pp. 69–73 (1998)
4. Buchandiran, G.: An exploratory study of indian science and technology publica-tion output. Libr. Philos. Pract. **658**, 1 (2011)
5. Ginde, G., et al.: ScientoBASE: a framework and model for computing scholas-tic indicators of non-local influence of journals via native data acquisition algo-rithms. Scientometrics **108**(3), 1479–1529 (2016). https://doi.org/10.1007/s11192-016-2006-2
6. Bora, K., Saha, S., Agrawal, S., Safonova, M., Routh, S., Narasimhamurthy, A.: CD-HPF: new habitability score via data analytic modeling. Astron. Comput. **17**, 129–43 (2016)
7. Saha S., et al.: Theoretical validation of potential habitability via analytical and boosted tree methods: An optimistic study on recently discovered exoplanets. arXiv preprint arXiv:1712.01040 (2017)
8. Sarasvathi, V., Iyengar, N.C.S.N., Saha, S.: QoS guaranteed intelligent routing using hybrid PSO-GA in wireless mesh networks. Cybern. Inf. Technol. **15**(1), 69–83 (2015)
9. Saha, S., Sarkar, J., Dwivedi, A., Dwivedi, N., Narasimhamurthy, A.M., Roy, R.: A novel revenue optimization model to address the operation and maintenance cost of a data center. J. Cloud Comput. **5**(1), 1–23 (2015). https://doi.org/10.1186/s13677-015-0050-8

10. Mohanchandra, K., Saha, S., Lingaraju, G.M.: EEG based brain computer interface for speech communication: principles and applications. In: Hassanien, A.E., Azar, A.T. (eds.) Brain-Computer Interfaces. ISRL, vol. 74, pp. 273–293. Springer, Cham (2015). https://doi.org/10.1007/978-3-319-10978-7_10

11. Jangid N., Saha S., Narasimhamurthy A., Mathur A.: Computing the Prestige of a journal: A Revised Multiple Linear Regression Approach. WCI-ACM Digital library (2015) (accepted)

12. Jangid, N., Saha, S., Gupta, S., Rao, M.J.: Ranking of journals in science and technology domain: a novel and computationally lightweight approach. In: IERI Procedia, pp. 57–62. Elsevier (2014)

13. Jenab, S.M.H., Nejati, A.: Evaluation of the scientific production of countries by a resource scaled two-dimensional approach. J. Scientometric Res. **3**(3), 95–103 (2014)

14. Das, A.K., Mishra, S.: Genesis Of altmetrics or article-level metrics for measuring efficacy of scholarly communications: current perspectives. J. Scientometric Res. **3**(2), 82–92 (2014)

15. Zupanc, G.K.H.: Impact beyond the impact factor. J. Compa. Physiol. A **200**(2), 113–116 (2013). https://doi.org/10.1007/s00359-013-0863-1

16. Moed, H.F.: Measuring contextual citation impact of scientific journals. J. Informetrics **4**(3), 265–277 (2010)

Stirling Numbers via Combinatorial Sums

Anwesh Bhattacharya[1]([⊠]) and Bivas Bhattacharya[2]

[1] Birla Institute of Technology and Science, Pilani, Rajasthan, India
`f2016590@pilani.bits-pilani.ac.in`
[2] PES University, Bangalore, India
`bivas@pes.edu`

Abstract. In this paper, we have derived a formula to find combinatorial sums of the type $\sum_{r=0}^{n} r^k \binom{n}{r}$ where $k \in \mathbb{N}$. The formula is conveniently expressed as a sum of terms multiplied by certain co-efficients. These co-efficients satisfy a recurrence relation, which is also derived in the process of finding the above sum. Upon solving the recurrence, these numbers turn out to be the Stirling Numbers of the first and second kind. Here on, it is trivial to prove the mutual inverse property of both these sequences of numbers due to linear algebra.

Keywords: Combinatorics · Stirling Numbers

1   Introduction

The inquiry into the matter started with the task to find the sum of $\sum_{r=0}^{n} r \binom{n}{r}$. Using the binomial expansion -

$$(1+x)^n = \sum_{r=0}^{n} x^r \binom{n}{r}$$
$$= \binom{n}{0} + x\binom{n}{1} + x^2\binom{n}{2} + \cdots + x^{n-1}\binom{n}{n-1} + x^n\binom{n}{n}$$

Thus, we have on differentiating -

$$\frac{d}{dx}(1+x)^n = \frac{d}{dx}\left[\binom{n}{0} + x\binom{n}{1} + x^2\binom{n}{2} + \cdots + x^{n-1}\binom{n}{n-1}\right]$$
$$+ x^n\binom{n}{n}$$
$$\implies n(1+x)^{n-1} = \binom{n}{1} + 2x\binom{n}{2} + 3x^2\binom{n}{3} + \cdots + nx^{n-1}\binom{n}{n} \tag{1}$$

© Springer Nature Singapore Pte Ltd. 2020
S. Saha et al. (Eds.): MMLA 2019, CCIS 1290, pp. 15–25, 2020.
https://doi.org/10.1007/978-981-33-6463-9_2

On setting x = 1 in Eq. 1, we get -

$$\binom{n}{1} + 2\binom{n}{2} + 3\binom{n}{3} + \cdots + n\binom{n}{n} = n2^{n-1}$$

$$\implies \sum_{r=0}^{n} r\binom{n}{r} = n2^{n-1} \tag{2}$$

which is our required answer. As a logical extension, one can ask what is the sum of $\sum_{r=0}^{n} r^2 \binom{n}{r}$.

On multiplying Eq. (1) with x and then differentiating, we obtain -

$$nx(1+x)^{n-1} = x\binom{n}{1} + 2x^2\binom{n}{2} + 3x^3\binom{n}{3} + \cdots$$
$$+ nx^n\binom{n}{n}$$

$$\implies \frac{d}{dx}nx(1+x)^{n-1} = \frac{d}{dx}\left[x\binom{n}{1} + 2x^2\binom{n}{2} + 3x^3\binom{n}{3} + \cdots\right.$$
$$\left. + nx^n\binom{n}{n}\right]$$

$$\implies n(1+x)^{n-1} + nx(n-1)(1+x)^{n-2} = \binom{n}{1} + 2^2 x\binom{n}{2} + 3^2 x^2\binom{n}{3} + \cdots$$
$$+ n^2 x^{n-1}\binom{n}{n} \tag{3}$$

Setting x = 1 in the above equation, we get -

$$\binom{n}{1} + 2^2\binom{n}{2} + 3^2\binom{n}{3} + \cdots + n^2\binom{n}{n} = n2^{n-1} + n(n-1)2^{n-2}$$

$$\implies \sum_{r=0}^{n} r^2\binom{n}{r} = n2^{n-1} + n(n-1)2^{n-2} \tag{4}$$

If we use the notation, $S_{nk} = \sum_{r=0}^{n} r^k\binom{n}{r}$, one can notice the pattern in Eqs. (2) and (4) and conjecture that -

$$S_{n3} = n2^{n-1} + n(n-1)2^{n-2} + n(n-1)(n-2)2^{n-3}$$

However, it turns out that it is not true because, on multiplying (3) with x and then differentiating -

$$n(1+x)^{n-1} + 3nx(n-1)(1+x)^{n-2} + nx^2(n-1)(n-2)(1+x)^{n-3}$$
$$= \binom{n}{1} + 2^3 x\binom{n}{2} + 3^3 x^2\binom{n}{3} + \cdots + n^3 x^{n-1}\binom{n}{n} \tag{5}$$

On setting $x = 1$ in (5), we get -

$$S_{n3} = n2^{n-1} + 3n(n-1)2^{n-2} + n(n-1)(n-2)2^{n-3} \tag{6}$$

As it can be seen, the $n(n-1)2^{n-2}$ term is padded with a co-efficient of 3. Hence, it does not suffice that the conjecture be so straightforward. Since the general method to find S_{nk} would require differentiating k times, we have to allow for co-efficients that multiply with the type of terms given above.

Thus, the task of finding the sum S_{nk} would be reduced to finding such co-efficients to multiply these terms with and there would be no need to differentiate k times explicitly.

2 A General Approach

If we denote $\prod_{i=0}^{k-1}(n-i) = (n)_k$ (A.K.A the falling factorial for $k \geq 1$), and $T_{ni} = (n)_i 2^{n-i}$, then -

$$S_{nk} = \sum_{i=1}^{k} a_{ki}(n)_i 2^{n-i} = \sum_{i=1}^{k} a_{ki}T_{ni} \qquad 1 \leq k \leq n \tag{7}$$

The a_{ki}'s are general co-efficients which are padded to T_{ni} terms. The sub-script a_{ki} has been chosen over a_{ni} in hindsight. The reason will be clear in the subsequent sections.

Note that $(n)_n = n!$ and for any $k > n$ $(n)_k = 0 \implies T_{nk} = 0$. We can also assume $(n)_0 = 1$. Hence it can also be stated -

$$S_{nk} = \sum_{i=1}^{k} a_{ki}(n)_i 2^{n-i} = \sum_{i=1}^{n} a_{ki}T_{ni} \qquad k > n \tag{8}$$

The sub-scripts in the formulae (28) and (8) have a subtlety. Once, the value of n for the problem is fixed, the T_{ni}'s are also fixed. The co-efficients a_{ki} however depend on the value of k in the original sum $\sum_{r=0}^{n} r^k \binom{n}{r}$.

Let's say we are evaluating S_{n3}. The final value of the sum is obtained by setting $x = 1$ in polynomial in x of (5) -

$$n(1+x)^{n-1} + 3nx(n-1)(1+x)^{n-2} + nx^2(n-1)(n-2)(1+x)^{n-3}$$

This can be generalised saying that the final series is obtained by setting $x = 1$ in a polynomial $S_{nk}^{(x)}$. In the previous case, this was $S_{n3}^{(x)}$. By analysing the patterns in the previous steps, if $S_{n4}^{(x)}$ had to be derived, we would multiply $S_{n3}^{(x)}$ with x and differentiate with respect to x. Generally -

$$S_{n(k+1)}^{(x)} = \frac{d}{dx}\left[xS_{nk}^{(x)}\right]$$

$$\implies S_{n(k+1)}^{(x)} = S_{nk}^{(x)} + x\frac{d}{dx}S_{nk}^{(x)} \tag{9}$$

where

$$S_{nk}^{(x)} = \sum_{i=1}^{k} a_{ki}(n)_i x^{i-1}(1+x)^{n-i} \qquad (10)$$

It is easy to see that every polynomial $S_{nk}^{(x)}$ is a polynomial in x of degree $n-1 \; \forall \; k \in \mathbb{N}$. This is because of recurrence relation (9) where the next polynomial of the sequence is obtained by first multiplying the previous polynomial by x and then differentiating. The boundary case of $S_{nn}^{(x)}$ can be considered as well. The last term in the sum will be -

$$S_{nn}^{(x)} = \cdots + a_{kn} n! x^{n-1}$$

To find $S_{n(n+1)}^{(x)}$, we need to multiply by x and then differentiate. Referring to Eq. (10), it can be seen that the last term of sum for $S_{n(n+1)}^{(x)}$ will again be $n! x^{n-1}$. However, it will be multiplied by a different co-efficient, namely $a_{(n+1)n}$. Hence, the clipping of the sum to n in (8) can be understood.

In other words, for $k > n$, the co-efficients a_{km} where $m > n$ may exist, but they are not required to evelute the original sum.

In general, the set of co-efficients $\{a_{(k+1)1}, a_{(k+1)2}, \cdots\}$ can be expressed in terms of $\{a_{k1}, a_{k2}, \cdots\}$ because of mixing of terms of equal powers and lowering of the exponent on differentiation in the derivation of the next polynomial of the sequence. The recurrence among polynomials (9) must naturally translate to a recurrence among co-efficients. In other words, we must be able to express the set $\{a_{(k+1)1}, a_{(k+1)2}, \cdots\}$ in terms of $\{a_{k1}, a_{k2}, \cdots\}$.

3   Obtaining the Recurrence Relation for a_{ki}

Let us assume $k \leq n$ and evelute all the terms in the sequence $\{a_{k1}, a_{k2}, \cdots, a_{kn}\}$.

The first term of $S_{nk}^{(x)}$ from (10) will always be $n(1+x)^{n-1}$. Due to the fact that $a_{11} = 1$ and (9), this term will always get carried on to the next polynomial sequence, and hence $a_{k1} = 1 \; \forall \; k \in \mathbb{N}$.

The last term is the result of differentiating k times and then multiplying by x. It is a newly generated term in the sequence and not carried over by previous polynomials of the sequence. Thus $a_{kk} = 1$ for $k \leq n$.

Consider the ith term of $S_{nk+1}^{(x)}$ and its co-efficient $a_{(k+1)i}$. Due to (9) and (10), the contributing terms to the ith term from $S_{nk}^{(x)}$ are -

$$a_{(k+1)i}(n)_i x^{i-1}(1+x)^{n-i} = a_{ki}(n)_i x^{i-1}(1+x)^{n-i}$$

$$+ a_{ki}(n)_i(1+x)^{n-i} x \frac{d}{dx} x^{i-1}$$

$$+ a_{k(i-1)}(n)_{i-1} x^{i-2} x \frac{d}{dx}(1+x)^{n-(i-1)}$$

$$\implies a_{(k+1)i}(n)_i x^{i-1}(1+x)^{n-i} = i a_{ki}(n)_i x^{i-1}(1+x)^{n-i}$$

$$+ a_{k(i-1)}(n)_i x^{i-1}(1+x)^{n-i} \qquad (11)$$

We can extract the recurrence relation by equating the co-efficients in (11) -

$$a_{(k+1)i} = ia_{ki} + a_{k(i-1)} \tag{12}$$

The co-efficients a_{ki} satisfy the same recurrence as that of the famous **Stirling Numbers of the Second Kind** and with the same base cases.

Hence these co-efficients *must* be the Stirling Numbers of the second kind. In their more common notation, they satisfy the recurrence -

$$\left\{ {k+1 \atop i} \right\} = i \left\{ {k \atop i} \right\} + \left\{ {k \atop i-1} \right\} \tag{13}$$

4   Triangle of Stirling Numbers of the Second Kind

It is well known that $\sum_{r=0}^{n} \binom{n}{r} = 2^n$. Here the exponent in r^k is $k = 0$.

Thus keeping in line with (28), we get -

$$S_{n0} = a_{00}T_{n0} \tag{14}$$

Since $(n)_0 = 1$ and $T_{n0} = 2^n$, from (14), it can be seen that -

$$a_{00} = 1 \tag{15}$$

For $k > 0$, the $T_{n0} = 2^n$ term in the expression for S_{nk} is missing. Hence -

$$a_{k0} = 0 \tag{16}$$

Since for $i > k$, the T_{ni} terms do not contribute to the sum as the upper limit of the summation in (28) is k. Thus we can safely define -

$$a_{ki} = 0 \qquad\qquad i > k \tag{17}$$

Thus we can combine (7) and (8) because of (15), (16) and (17) to generalise

-

$$S_{nk} = \sum_{i=1}^{n} a_{ki}T_{ni} \qquad\qquad k \in \mathbb{N} \tag{18}$$

We can display these numbers in a triangular fashion (for $0 \le k \le 10$) by building the recurrence -

5   Verifying the Formula for S_{nk}

We can confirm the validity of the method for two examples - one with $k \le n$ and another with $k > n$ (Table 1).

Table 1. Stirling Numbers of the Second Kind

i		0	1	2	3	4	5	6	7
k	0	1	0	0	0	0	0	0	0
	1	0	1	0	0	0	0	0	0
	2	0	1	1	0	0	0	0	0
	3	0	1	3	1	0	0	0	0
	4	0	1	7	6	1	0	0	0
	5	0	1	15	25	10	1	0	0
	6	0	1	31	90	65	15	1	0
	7	0	1	63	301	350	140	21	1

5.1   Case 1 - $n = 5, k = 4$

$$\sum_{r=0}^{5} r^4 \binom{5}{r} = 0^4 \binom{5}{0} + 1^4 \binom{5}{1} + 2^4 \binom{5}{2} + 3^4 \binom{5}{3} + 4^4 \binom{5}{4} + 5^4 \binom{5}{5}$$
$$= 0 + (1 \times 5) + (16 \times 10) + (81 \times 10) + (256 \times 5) + (625 \times 1)$$
$$= 2880$$

From (18), we have -

$$S_{54} = \sum_{i=0}^{5} a_{4i} T_{5i}$$
$$= a_{40} T_{50} + a_{41} T_{51} + a_{42} T_{52} + a_{43} T_{53} + a_{44} T_{54} + a_{45} T_{55}$$
$$= 0 + (1 \times 80) + (7 \times 160) + (6 \times 240) + (1 \times 240) + 0$$
$$= 2880$$

Hence it matches!

5.2   Case 2 - $n = 3, k = 6$

$$\sum_{r=0}^{3} r^6 \binom{3}{r} = 0^6 \binom{3}{0} + 1^6 \binom{3}{1} + 2^6 \binom{3}{2} + 3^6 \binom{3}{3}$$
$$= 0 + (1 \times 3) + (64 \times 3) + (729 \times 1)$$
$$= 924$$

From (18), we have -

$$S_{36} = \sum_{i=0}^{3} a_{6i} T_{3i}$$

$$= a_{60} T_{30} + a_{61} T_{31} + a_{62} T_{32} + a_{63} T_{33}$$

$$= 0 + (1 \times 12) + (31 \times 12) + (90 \times 6)$$

$$= 924$$

It matches too!

6 Another Approach

We shall derive the inverse relation i.e. T_{nk} in as a linear sum of S_{nk}'s. From this point on, we shall asume strictly $k \leq n$

$$(1+x)^n = \sum_{r=0}^{n} x^r \binom{n}{r}$$

Differentiating k times -

$$\left[\prod_{i=0}^{k-1} (n-i) \right] (1+x)^{n-k} = \sum_{r=k}^{n} \left[\prod_{i=0}^{k-1} (r-i) \right] x^{r-k} \binom{n}{r} \qquad (19)$$

The product on the LHS is just the falling factorial. One can expand the product on the RHS as (b_{ki}'s are general co-efficients) -

$$\prod_{i=0}^{k-1} (r-i) = \sum_{i=1}^{k} b_{ki} r^i \qquad (20)$$

Plugging in (20) in (19) and multiplying both sides by x^k -

$$(n)_k x^k (1+x)^{n-k} = \sum_{r=k}^{n} \sum_{i=1}^{k} b_{ki} r^i x^r \binom{n}{r}$$

$$= \sum_{r=0}^{n} \sum_{i=1}^{k} b_{ki} r^i x^r \binom{n}{r} - \sum_{r=0}^{k-1} \sum_{i=1}^{k} b_{ki} r^i x^r \binom{n}{r}$$

$$= \sum_{i=1}^{k} b_{ki} \sum_{r=0}^{n} r^i x^r \binom{n}{r} - \sum_{r=0}^{k-1} x^r \binom{n}{r} \left[\sum_{i=1}^{k} b_{ki} r^i \right]$$

$$= \sum_{i=1}^{k} b_{ki} \sum_{r=0}^{n} r^i x^r \binom{n}{r} - \sum_{r=0}^{k-1} x^r \binom{n}{r} \left[\prod_{i=0}^{k-1} (r-i) \right] \qquad (21)$$

Plugging in $x = 1$ in (21) and using our defined notations -

$$(n)_k 2^{n-k} = \sum_{i=1}^{k} b_{ki} \left[\sum_{r=0}^{n} r^i \binom{n}{r} \right] - \sum_{r=0}^{k-1} \binom{n}{r} \left[\prod_{i=0}^{k-1} (r-i) \right]$$

$$\implies T_{nk} = \sum_{i=1}^{k} b_{ki} S_{ni} - \sum_{r=0}^{k-1} \binom{n}{r} \left[\prod_{i=0}^{k-1} (r-i) \right] \tag{22}$$

In the summation indexed by r on the RHS of (22), r can only take values from $\{0, 1, \cdots, k-1\}$. The product vanishes for every value of r as i indexes from 0 to $k-1$. Hence the last summation is identically zero. Ultimately -

$$T_{nk} = \sum_{i=1}^{k} b_{ki} S_{ni} \tag{23}$$

The subscripts of the terms here follow the same pattern as that of (28) but due to different reasons. The b terms are indexed by k first because b_{ki} represents the coefficient of r^i in $\prod_{i=0}^{k-1}(r-i)$ and the maximum power in this product is r^k, and hence the indexing.

One can also observe that (23) is simply the inverse relationship of $S_{nk} = \sum_{i=1}^{k} a_{ki} T_{ni}$.

We can expect to derive, in a similar fashion, a recurrence relation for $\{b_{(k+1)1}, b_{(k+1)2}, \cdots\}$ in terms of $\{b_{k1}, b_{k2}, \cdots\}$.

7   Obtaining the Recurrence Relation for b_{ki}

From the definition of b_{ki}, it can be seen that the co-efficient of the lowest power is $b_{k1} = \prod_{i=1}^{k-1} (-1)^i i = (-1)^{k-1}(k-1)!$. Also, the co-efficient of the highest power is $b_{kk} = 1$.

We have established the base cases and can continue to establish the recurrence relation. From (20), we have -

$$\prod_{i=0}^{k} (r-i) = \sum_{i=1}^{k+1} b_{(k+1)i} r^i$$

$$\implies \sum_{i=1}^{k+1} b_{(k+1)i} r^i = (r-k) \left[\prod_{i=0}^{k-1} (r-i) \right]$$

$$= (r-k) \left[\sum_{i=1}^{k} b_{ki} r^i \right]$$

$$= \sum_{i=1}^{k} b_{ki} r^{i+1} - \sum_{i=1}^{k} k b_{ki} r^i$$

$$\sum_{i=1}^{k+1} b_{(k+1)i} r^i = \sum_{i=2}^{k-1} b_{k(i-1)} r^i - \sum_{i=1}^{k} k b_{ki} r^i \qquad (24)$$

The summation limits only differ in the boundary cases which have already been derived. Hence, from (24), we can state a recurrence relation -

$$b_{(k+1)i} = b_{k(i-1)} - k b_{ki} \qquad (25)$$

Again, we discover that b_{ki} satisfy the same recurrence as that of **Signed Stirling Numbers of the First Kind** and with the same base cases. As follows from the previous argument, these *must* be the Signed Stirling Numbers of the First Kind. They are called *signed* as some of these numbers are negative.

$$\begin{bmatrix} k+1 \\ i \end{bmatrix} = \begin{bmatrix} k \\ i-1 \end{bmatrix} - k \begin{bmatrix} k \\ i \end{bmatrix} \qquad (26)$$

8 Triangle of Signed Stirling Numbers of the First Kind

Setting the index $k = 0$ in (23) and letting i run from 0, we get -

$$T_{n0} = b_{00} S_{n0} \qquad (27)$$

Since $T_{n0} = S_{n0} = 2^n$, we can set $b_{00} = 1$

Also, by the recurrence relation (25), we can see that $b_{k1} = (-1)^k (k-1)!$ iff the recurrence satisfied by $i = 1$ is $b_{(k+1)1} = -k b_{k1}$. Thus, $b_{k0} =$ has to be satisfied. In the later sections, this can also shown to be true by linear algebra (Table 2).
We can again display these numbers in a triangular fashion -

Table 2. Signed Stirling Numbers of the First Kind

i		0	1	2	3	4	5	6	7
k	0	1	0	0	0	0	0	0	0
	1	0	1	0	0	0	0	0	0
	2	0	−1	1	0	0	0	0	0
	3	0	2	−3	1	0	0	0	0
	4	0	−6	11	−6	1	0	0	0
	5	0	24	−50	35	−10	1	0	0
	6	0	−120	274	−225	85	−15	1	0
	7	0	720	−1764	1624	−735	175	−21	1

9   Verifying the Formula for T_{nk}

It has already been stated that $k \leq n$. Hence, we shall verify the formula (23) for two cases:

9.1   Case 1 - $n = 5, k = 3$

$$T_{53} = 240$$
$$S_{51} = 80$$
$$S_{52} = 240$$
$$S_{53} = 800$$
$$b_{31}S_{51} + b_{32}S_{52} + b_{33}S_{53} = (2 \times 80) - (3 \times 240) + (1 \times 800)$$
$$= 240$$

Thus it mactches as expected.

9.2   Case 2 - $n = 6, k = 4$

$$T_{64} = 1440$$
$$S_{61} = 192$$
$$S_{62} = 672$$
$$S_{63} = 2592$$
$$S_{64} = 10752$$
$$b_{41}S_{61} + b_{42}S_{62} + b_{43}S_{63} + b_{44}S_{64} = (-6 \times 192) + (11 \times 672) - (6 \times 2592) + (1 \times 10752)$$
$$= 1440$$

This too matches as expected!

10   Proving the Inverse Nature of the Two Sequences

We shall concern ourselves only with a square sub-section of the table of Stirling numbers (i.e. $1 \leq k \leq n$).

Observe the formula for $T_{nk} = (n)_k 2^{n-k} = n \times (n - 1) \times \cdots \times (n - k + 1)2^{n-k}$. This can be looked at as a polynomial in n of degree k. The set $\{T_{n1}, T_{n2}, \cdots, T_{nn}\}$ essentially contains polynomials in n of degree $1, 2, \cdots$ upto n. Hence, it can act as a basis for the vector space denoted by $span\left(\{n, n^2, n^3, \cdots, n^n\}\right)$.

Similarly, the set $\{S_{n1}, S_{n2}, \cdots, S_{nn}\}$ is also a set of polynomials in n and (28) basically represents a linear transformation in the co-ordinatization of the $\{T_{nk}\}$ basis state.

By using (28) and (23), we get -

$$T_{nk} = \sum_{i=1}^{k} b_{ki} S_{ni}$$

$$= \sum_{i=1}^{k} b_{ki} \left(\sum_{l=1}^{i} a_{il} T_{nl} \right)$$

$$= \sum_{i=1}^{k} \sum_{l=1}^{i} b_{ki} a_{il} T_{nl}$$

The inner sum runs from $l = 1$ to i. We note that $i \leq k$ due to the outer sum and it does not make a difference to change the upper limit of the inner sum to k. Since $\{T_{nk}\}$ represents a basis state, we must have -

$$\sum_{i=1}^{k} \sum_{l=1}^{k} b_{ki} a_{il} T_{nl} = \sum_{l=1}^{k} \delta_{kl} T_{nl}$$

$$= T_{nk}$$

With the Einstein summation convention, we can say $b_{ki} a_{ij} = \delta_{kj}$ and that the square matrices represented by a square-section of the two tables of the Stirling Numbers are inverse with respect to each other.

11 Verifying the Inverse Property

Let us verify our result with a 6×6 sub-matrix of the two tables.

$$B = \begin{bmatrix} 1 & 0 & 0 & 0 & 0 & 0 \\ -1 & 1 & 0 & 0 & 0 & 0 \\ 2 & -3 & 1 & 0 & 0 & 0 \\ -6 & 11 & -6 & 1 & 0 & 0 \\ 24 & -50 & 35 & -10 & 1 & 0 \\ -120 & 274 & -225 & 85 & -15 & 1 \end{bmatrix} \qquad A = \begin{bmatrix} 1 & 0 & 0 & 0 & 0 & 0 \\ 1 & 1 & 0 & 0 & 0 & 0 \\ 1 & 3 & 1 & 0 & 0 & 0 \\ 1 & 7 & 6 & 1 & 0 & 0 \\ 1 & 15 & 25 & 10 & 1 & 0 \\ 1 & 31 & 90 & 65 & 15 & 1 \end{bmatrix}$$

Carrying out the matrix multiplication $B \times A$ results in the identity matrix I_6. We could have included the 00th index term in the matrices and it is trivial to confirm that the result would have remained the same.

References

1. Triangle of Signed Stirling Numbers of The First Kind, OEIS Foundation Inc. (2019), The On-Line Encyclopedia of Integer Sequences. https://oeis.org/A048994
2. Triangle of Stirling Numbers of The Second Kind, OEIS Foundation Inc. (2019), The On-Line Encyclopedia of Integer Sequences. https://oeis.org/A008277

Random Subspace Combined LDA Based Machine Learning Model for OSCC Classifier

Archana Nawandhar[1](✉), Navin Kumar[2], and Lakshmi Yamujala[3]

[1] CMRIT, ECE, Bangalore 590018, India
archusarda@rediffmail.com
[2] Amrita School of Engineering, Bengaluru Amrita Vishwa Vidyapeetham, Bangalore 560035, India
[3] Centre for Development of Telematics (C-DOT), Bangalore 560100, India

Abstract. Oral squamous cell carcinoma (OSCC) remains a major death causing oral cancer in developing countries. In recent years, tremendous development in medical imaging devices made microscopic colour images of biopsy samples available to the researchers. Image processing and machine learning techniques can be used to develop automatic cancer grading mechanism. In this work, automatic OSCC classifier using Linear Discriminant Analysis combined with Random Subspace is developed and analyzed. The proposed classifiers automatically classifies the input image in one of the four categories, namely: Normal, Grade-I, II or III. Total 83 colour and texture features are computed from the 100 Haemotoxylin and Eosin (H&E) stained images of oral mucosa. The overall accuracy of the proposed classifier is 93.5% with sensitivity and specificity of 0.89 and 0.95 respectively.

Keywords: LDA · Random subspace · OSCC

1 Introduction

Oral cancer is one of the life-threatening ailments in the world. Among all the types of oral cancers, oral squamous cell carcinoma (OSCC) is most common form of cancer usually found in people with tobacco chewing and smoking habits [1]. Traditional way of cancer diagnosis is highly dependent upon clinico-pathological acumen of the diagnosing experts [2, 3]. Furthermore, the cancer detection and grading is highly dependent upon experts and their experience. With the progressions in the medical imaging equipment, digital images of biopsy samples are now easily available. Using image processing techniques and machine learning algorithms, it's possible to develop an efficient automatic image classifier for detecting and classifying the microscopic images into normal and malignant lesion with different cancer grades. Computer assisted identification and grading of the cancer can be taken as an important compliment to the pathologist's diagnosis. In last decade many attempts have been made by various researchers to develop such a system for various types of cancer. In [4] authors have suggested image analysis for detection of OSCC using Tissue Microarray (TMA). Some researchers [5] used cytology images for detecting oral cancer. Authors in [6] used image processing techniques

© Springer Nature Singapore Pte Ltd. 2020
S. Saha et al. (Eds.): MMLA 2019, CCIS 1290, pp. 26–40, 2020.
https://doi.org/10.1007/978-981-33-6463-9_3

to detect OSCC using Computed Tomography (CT) images. In this work we focus on automatic detection and classification of OSCC which originates in stratified epithelium of oral cavity using Hematoxylin and Eosin (H&E) stained microscopic images using machine learning technique.

Common procedure of developing classifier involves extracting discriminating features from the input images and feeding these features to the developed classifiers to get the final classification. Linear discriminant analysis (LDA) is one of the recognized classifiers explored by multiple authors to develop oral cancer classifiers. In [7], authors have used complexity of fractal geometry at local and global level to classify the OSCC using LDA. Their proposed system exhibit sensitivity of 63% and specificity of 67%. The authors of [8] explored set of texture features to classify the oral sub-mucous fibrosis (OSF) in normal and abnormal using LDA. The classifier demonstrated 88.38% accuracy. In [9], authors have used morphological features of segmented cells to identify presence of dysplasia in epithelial layer of oral cavity which can be indication of the presence of cancer. Their proposed classifier used LDA and overall accuracy seen was 46%. Staining of the biopsy lesion highlights various cellular regions by different colours and intensity. These colour and intensity features has potential to detect abnormalities in the lesion and the use of colour features with LDA for detection and classification of OSCC is yet to be explored.

In this study, a novel technique of retaining the colour information while preprocessing the microscopic images is developed and used. Further LDA classifier combined with random subspace technique with colour and texture features is trained, tested and analyzed. These features are extracted from the microscopic images of stratified squamous epithelium (SSE). The set of input images include normal lesion, well differentiated OSCC, moderately-differentiated and poorly-differentiated OSCC images [10]. The proposed classifier exhibits inclusive accuracy of 93.5% with specificity of 0.95 and sensitivity of 0.89. The performance of suggested classifier LDA with RS using colour and texture features is promising to be used as OSCC classifier.

The rest of the paper include the following sections: Sect. 2 describes the characteristics of input images and proposed method. Section 3 presents the feature extraction technique and Sect. 4 is about classifier. The performance measures used are described in Sect. 5. Results and discussion are presented in Sect. 6 followed by conclusion in Sect. 7.

2 Input Images and Methods

In this section details about the dataset preparation and peculiar characteristics of the input images are discussed.

2.1 Input Dataset

The dataset used for this classifier are provided by healthcare global enterprises limited (HCG) hospitals, Bangalore-Karnataka, India. Olympus CX31 microscopes is used to capture all the images [11]. These microscopic images are RGB-Coloured and of H&E-stained biopsy lesion of SSE. This image dataset consists of images under four categories:

Normal sample-40, Well-differentiated (WD)-20, moderately-differentiated (MD)-30 and poorly-differentiated (PD)-10. Total 100 images captured with 10X magnification and resolution of 960 X 720. Figure 1 (a) to (d) shows sample images of Normal, WD, MD and PD OSCC.

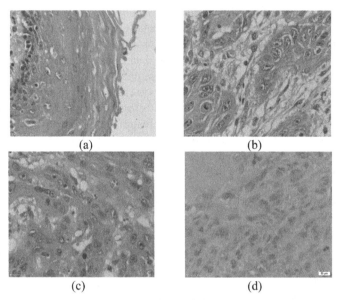

Fig. 1. Sample H&E-stained microscopic images of stratified epithelium of oral cavity (a) normal, (b) well differentiated, (c) moderately differentiated, (d) poorly differentiated (Color figure online)

OSCC is a multi-phase progression causing the disruption of the normal controlling pathways that regulate basic cellular functions including cell division, differentiation, and cell death. This disturbance leads to abnormal increase in number of cell surface, shape and size of cell nucleus, disturbance in regular structure of connective tissue. The genetic action fails in overturning the tumor spread resulting into abnormal cell phenotype which escalates cell proliferation. It also damages cell cohesion and can infiltrate local tissue and spread to distant sites [10]. H&E-taining highlights the cell nucleus by staining it in purple-blue colour and cytoplasm takes pink colour. The colour intensity varies in abnormal tissue from the normal one. Obviously, the colour details can be used as abnormality indicating features. Similarly, the arrangement of various cellular structures gets disturbed and become irregular in malignant lesion which is regular in non-malignant. Thus, texture features are important and added to the feature set.

2.2 Proposed Classifier

Figure 2 shows flow diagram of proposed OSCC classifier using LDA combined with random subspace (RS). The objective of the classifier is to classify the microscopic images of SSE layer in Normal/ WD- /MD-/PD-OSCC categories. The input images are in RGB color model format. These are converted into CIE L*a*b* and HSV colour space. After that color features are calculated from all three color modelled images. For extracting texture features L-channel of L*a*b* color converted images. Finally feature sampling is performed over the feature set to create feature subset as per the RS technique. All these feature subsets are used to train multiple LDA classifiers and final prediction is carried out by majority voting.

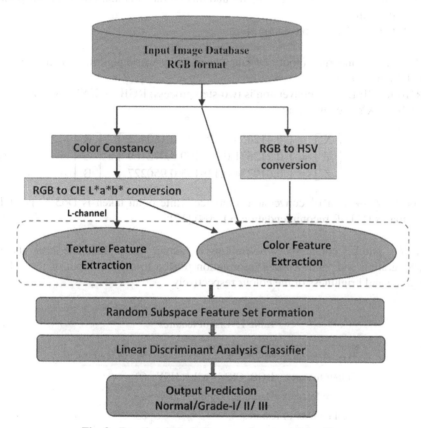

Fig. 2. Functional block diagram of proposed classifier

3   Feature Extraction

As mentioned in Sect. 2.1, colour features and texture features are calculated from the input images. Feature set consist of total 83 features as shown in Table 1.

3.1   Colour Features

The available dataset consist of RGB colour model images. These images converted into CIE L*a*b* colour space and HSV colour model. CIE L*a*b* colour space is perceptually linear and analogous to how human perceive the colours. Second advantage of this conversion is, the L-plane can be used as grayscale image since it carries only intensity information. Colour information will remain unchanged and retained in *a and *b plane. For RGB to L*a*b* conversion reference illumination value of white light is required representing the surrounding illuminance. As mentioned before microscopic images are captured in artificial light in the laboratories due to which it's difficult to decide the white point value. To work around this problem, colour constancy is applied to all the input images.

RGB to CIE L*a*b* conversion steps:

1) RGB image undergo colour constancy using Grey-world algorithm then converted to L*a*b* colours.
2) RGB to CIE L*a*b* conversion is two-step process: RGB → XYZ → L*a*b*
3) RGB → XYZ conversion:

$$\begin{bmatrix} X \\ Y \\ Z \end{bmatrix} = \begin{bmatrix} 0.412453 & 0.35758 & 0.180423 \\ 0.212671 & 0.715160 & 0.072169 \\ 0.019334 & 0.119193 & 0.950227 \end{bmatrix} * \begin{bmatrix} R \\ G \\ B \end{bmatrix} \tag{1}$$

4) For XYZ → L*a*b* conversion reference white point taken is D65 as per CIE standards (ITU-R recommendation BT.709) [12]

HSV is another colour space resembling to human vision [13]. The three planes; Hue, Saturation and Value carries information related to colour, depth of the colour (Grayness) and brightness of the colour respectively.

Table 1. List of features

Type of features		No. of features
Colour features	RGB, CIE-L*a*b*, HSV	60
Texture features	Haralick, Tamura, Laws	23
Total number of features		83

RGB to HSV conversion steps:
RGB values are divided by 255 to normalize the range from 0 → 255 to 0 → 1

1.

$$R' = \frac{R}{255}, \ G' = \frac{G}{255}, \ B' = \frac{B}{255} \tag{2}$$

2.

$$C_{\max} = \max(R', G', B'), \ C\_\min = \min(R', G', B'),$$

3.

$$\Delta = C_{max} - C_{min} \tag{3}$$

4.

$$\text{Hue, } H = \begin{cases} 0^0 & \Delta = 0 \\ 60^0 X \left(\frac{G-B}{\Delta} mod 6 \right), & Cmax = R \\ 60^0 X \left(\frac{B-R}{\Delta} + 2 \right), & Cmax = G \\ 60^0 X \left(\frac{R-G}{\Delta} + 4 \right), & Cmax = B \end{cases} \tag{4}$$

5.

$$\text{Saturation, } S = \begin{cases} 0, & C_{max} = 0 \\ \frac{\Delta}{C_{max}}, & C_{max} \neq 0 \end{cases} \tag{5}$$

6.

$$\text{Value, } V = C_{max}$$

After this conversion, seven colour features from all the colour planes of RGB model, L*a*b* and HSV colour space converted images are calculated. These include three statistical features: mean, std. deviation, skewness and four Haralick features: contrast, energy, homogeneity and correlation.

3.2 Texture Features

Haralick texture features [14], Tamura texture features [15] and Laws texture features [16] are extracted from grey-scale images. L-channel of L*a*b* colour space converted images is taken as grey image.

Haralick Texture Features. Haralick texture features are computed using gray level co-occurrence matrix (GLCM). These are well known and widely used texture features. Total 14 features were calculated from GLCM matrix prepared from grey-scale image. These include energy, contrast, correlation, variance, homogeneity, sum of average, sum of variance, sum of entropy, entropy, diff. of variance, diff. of entropy, correlation-I, correlation-II, maximal correlation coefficient.

Tamura Texture Features. Tamura features are like texture features perceived by human beings from the images. Three main Tamura features namely coarseness, contrast and directionality are computed from grey image. Coarseness captures local variations in the intensity levels with respect to texture primitive elements called as texels. The contrast measures the global variations in the intensity level and its bias towards the black or white in the image. It is computed using variance, σ^2 and kurtosis, α_4 as $F_{con} = \frac{\sigma}{\alpha_4^n}$ where, $\alpha_4 = \mu_4/\sigma^4$; $\mu_4 = \sum_q (q - m)^4 \Pr(q|g)$; $\sigma^2 = \sum_q (q - m)^2 \Pr(q|g)$ and $m = $ mean gray level g is gray level. The value of $n = 0.25$ is considered as best discriminating textures. Orientation information is captured by the degree of directionality. It is measured using Sobel edge detector which computes first degree derivative of the image in x and y direction in terms of edge strength, $Es(x, y) = 0.5(|\Delta x(x, y)| + |\Delta y(x, y)|)$ and directional angle, $Da(x, y) = \tan^{-1}\left(\frac{\Delta y(x, y)}{\Delta x(x, y)}\right)$. Finally, regularity of oriented local edges against their directional angles gives the degree of directionality.

Laws Texture Features. The second set of texture features are Laws texture energy features. Six convolution masks are constructed from three Laws vectors: Level (L5) = [1 4 6 4 1], Edge (E5) = [−1 −2 0 2 1], and Spot (S5) = [−1 0 2 0 −1]. The details of the features captured is listed in Table 2. Five neighborhood region is considered for these masks. The gray image is convolved with these masks to extract six texture features.

Table 2. Laws texture features description

Convolution masks	Feature description
$L5^T L5$	Detects Grey level intensity in vertical and horizontal direction
$E5^T E5$	Detects edges in vertical and horizontal direction
$S5^T S5$	Detection of spots in vertical and horizontal direction
$[(L5^T E5) + (E5^T L5)]/2$	Average texture energy with respect to grey level intensity and edge in vertical and horizontal direction
$[(E5^T S5) + (S5^T E5)]/2$	Average texture energy with respect to spot and edge in vertical and horizontal direction
$[(L5^T S5) + (S5^T L5)]/2$	Average texture energy with respect to grey level intensity and edge vertical and horizontal direction

4 Random Subspace Combined Linear Discriminant Analysis Classifier

LDA is originally suggested by Fisher [17] for two class classification problem. Later C R Rao [18] extended the idea for multiclass classification problems. The idea behind LDA is to construct discriminant functions which are the optimum linear combination of features. These functions are then used as new latent variables in place of original feature predictors for classification.

4.1 Multiclass LDA

Consider D-dimensional training dataset, $T_r = \{(x_1^D, y_1), (x_2^D, y_2), \ldots, (x_i^D, y_i)..(x_N^D, y_N)\}$ of class $y = \{1, 2, \ldots, C\}$. Within-class scatter matrix S_w and between-class scatter matrix S_b are computed as given in Eq. 6 and 7.

$$S_w = \sum_i^c \sum_{x \in k_i} (x - \overline{x_i})(x - \overline{x_i})' \tag{6}$$

$$S_b = \sum_{i=1}^{C} m_i (\overline{x_i} - \overline{x})(\overline{x_i} - \overline{x})' \tag{7}$$

where, m_i = sample size of class i; $\overline{x}$ is mean vector given as $\overline{x} = \frac{1}{m} \sum_i m_i \overline{x_i}$ and $\overline{x_i}$ is the mean for each class. Linear transformation ϕ is obtained by solving generalized eigenvalue problem, $S_w \phi = \lambda S_b \phi$. Then in transformed space classification is performed using Euclidean distance metric given as: $d(x, \overline{x}) = \sqrt{\sum_i (x_i - \overline{x_i})^2}$. Though multiclass LDA is easy to implement and a promising classifier, it shows poor performance when number of dataset is smaller than number of predictors or features.

Often medical images availability is limited. A lot of parameters and factors take part into computer aided diagnosis due to which requirement of number of feature to be extracted from these images eventually larger than the number of samples. In current scenario the availability of images of each category is in the range of 10 to 50 whereas the feature set consists of 83 features.

RS technique is the solution to such situations. In RS, the training data is altered with respect to the feature space and not the sample data. Multiple subsets of features are prepared. The classifier is trained over all these subsets and finally combined decision is taken to declare the classification result.

4.2 Random Subspace Technique

Continuing with the same terminology for training set, $T_r = \{x_1^D, x_2^D, \ldots, x_i, \ldots x_N^D\}$; where N is number of samples and each sample is a D-dimensional vector. This means D is number of features. In RS, m randomly selected subsets of features each comprising p features such that $p < D$ are prepared. Thus, the classifier will have m feature subsets of dimension p as its training data. So the modified training samples will be $\widetilde{T}_r = \{\widetilde{T}_{r1}^p, \widetilde{T}_{r2}^p, \ldots, \widetilde{T}_{ri}^p, \ldots \widetilde{T}_{rm}^p\}$, where $\widetilde{T}_{ri}^p = \{x_1^p, x_2^p, \ldots, x_i^p, \ldots x_N^p\}$ is modified set of training data comprising p features. Such m sets are prepared. Most important part in this scheme is to decide correct values for number of feature subsets to be prepared and number of features to be sampled for each subset. Tuning of these parameter is carried out using grid search technique and then optimum values are selected.

Once the modified training data is prepared then LDA is trained using all these subsets and final classification is carried out using majority voting.

5 Performance Measures

The performance of proposed RS combined LDA OSCC classifier is evaluated by well-known performance parameters. These include receiver operating characteristic curve (ROC), accuracy, sensitivity, specificity, F-score and Mathew's correlation coefficient (MCC). True Positive (TP), True Negative (TN), False Positive (FP), and False Negative (FN) for each class is calculated using 4X4 confusion matrix. These are then utilized to calculate performance parameters. Formulas and standard definitions of performance parameters are briefly described below.

5.1 Accuracy

Accuracy of the classification technique depends on the number of correctly classified samples (i.e. TP and TN) [19]. It is calculated as given below:

$$Accuracy = \frac{TP + TN}{N} * 100; \ N = total\ number\ of\ microscopic\ images. \tag{8}$$

5.2 Sensitivity

It is the measure of the proportion of positive samples which are correctly classified [19]. It is calculated as given below:

$$Sensitivity = \frac{TP}{TP + FN} \tag{9}$$

5.3 Specificity

It is a measure of the proportion of negative samples that are correctly classified [19] and can be calculated as given below:

$$Specificity = \frac{TN}{TN + FP} \tag{10}$$

5.4 F-Score

F-score is a weighted average of precision and recall; and can be calculated as:

$$Precision: \frac{TP}{TP + FP}, Recall = \frac{TP}{TP + FN} \tag{11}$$

$$F - score = 2 * \frac{Precision * Recall}{Precision + Recall} \tag{12}$$

5.5 Matthew's Correlation Coefficient (MCC)

MCC is the measure of the eminence of binary class classification. For multiclass classification problem, overall MCC calculated using 'Macro-averaging' scheme as described below.

Assume $i = \{1, 2, 3, 4\}$ are four classes. Samples will be classified as of one of the class. Then 4×4 confusion matrix is used to calculate TP, TN, FP and FN as:

$$TP = \sum_i TP_i, \ TN = \sum_i TN_i, \ FP = \sum_i FP_i, \ FN = \sum_i FN_i \tag{13}$$

Using these values inclusive MCC is calculated as shown below

$$MCC = \frac{TP * TN - FP * FN}{\sqrt{((TP + FN)(TP + FP)(TN + FN)(TN + FP))}} \tag{14}$$

For class wise analysis, 'Macro averaging' scheme is implemented. 'One vs All' strategy is used.

5.6 Receiver Operating Characteristics Curve (ROC) and Area Under Curve (AUC)

ROC is a plot of the false positive rate (x-axis) also referred as specificity versus the true positive rate (y-axis) also referred as sensitivity for a number of variable threshold values between 0.0 and 1.0 [20]. Details of sensitivity and specificity is already mentioned in Sects. 5.2 and 5.3 respectively. Classifier with 100% correct classification ability has a ROC curve that passes through the upper left corner with 100% sensitivity and 100% specificity. Thus, the closer the ROC curve is to the upper left corner of the plot, the higher the total precision of the classifier. AUC is area under ROC curve represents the accuracy of the classifier. If it's in the range of 90–100 than then classifier is taken as excellent.

6 Results and Discussion

Stepwise working of proposed classifier is presented in Fig. 2 in Sect. 2.2. Aim of the proposed classifier is to classify the microscopic images of SSE (H&E-stained) in one of the four classes, Normal/WD/MD/PD OSCC. After feature set is computed as described in Sect. 3, sampling of feature set is performed as per the RS-scheme explained in Sect. 4.2. Figure 3 shows sample images of colour converted images. It is an important step of feature extraction procedure explained in Sect. 3.

6.1 Optimizing the Classifier for Key Parameters

Two levels of hyper-parameter optimization is required for the proposed RS combined LDA classifier. First is the regularization hyper-parameters for LDA. Secondly as per RS scheme, number of feature subsets (m) and number of features (p) in each subset needs to be selected such that performance of the classifier is optimized.

The optimization of regularization parameters δ and γ is carried out using Bayesian optimization technique subject to classification loss. The optimization process is shown graphically in Fig. 4. The optimum value for $\delta = 0.0483$ and $\gamma = 0.1387$ are selected with predicted classification error $Er = 0.084$.

Once the optimum values of δ and γ are calculated, LDA is further trained for best suitable values of number of features per feature subset and number of subsets. To perform this selection, grid search method is applied. LDA is trained using 10 feature subsets randomly selected from the feature set ranging from 3 to 83 and 10-fold classification error is calculated. Selection of minimum number of feature subsets is shown graphically in Fig. 5. It can be observed that any value between 18 to 30 subsets provides minimum classification accuracy of 0.08 which is lesser than estimated error value for the classifier. Finally, the subset size depicting minimum classification error was selected as final feature subset size. It is 56 for this case as shown in Fig. 6. So, 25 feature subsets were prepared each consisting of 56 features sampled from 83 features.

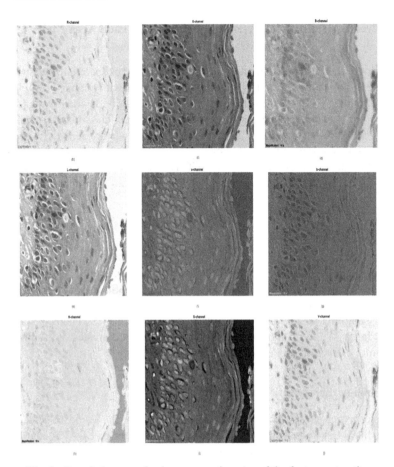

Fig. 3. Sample images of colour conversion step of the feature extraction

6.2 Performance Analysis of the Classifier

RS combined LDA is trained and tested with 10-fold cross validation. Table 3 shows the performance statistics of the classifier in the form of (mean ± std. dev.). The classifier exhibits best performance PD OSCC class than rest. Overall accuracy for the classifier is 93.5% with specificity, sensitivity, F-score and MCC of 0.95, 0.89, 0.89 and 0.84 respectively. Figure 6 shows the ROC curve for the classifier. Area under curve (AUC) for the proposed classifier is 96.92 indicates proposed classifier falls under excellent category (Fig. 7).

6.3 Comparative Analysis

Table 4 summarizes the comparative analysis of proposed method with the existent LDA classifier used for oral cancer classification. It can be observed that among all the type of features proposed by other researchers, current set of features provides superior performance for four class classification using RS.

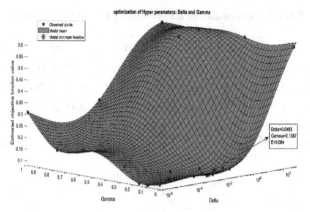

Fig. 4. Optimization of δ and γ

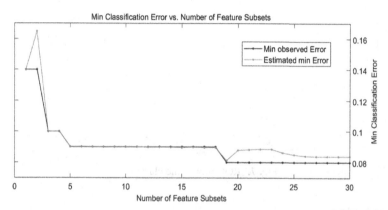

Fig. 5. Random subspace parameter optimization: optimum value for no. of subsets

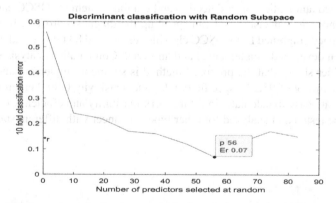

Fig. 6. Random subspace parameter optimization: optimum value for features/subset

Table 3. Performance of RS combined LDA OSCC classifier

	Accuracy (%)	Sensitivity	Specificity	F-score	MCC
Normal	90.0 ± 0.63	0.89 ± 0.02	0.90 ± 0.01	0.87 ± 0.02	0.79 ± 0.02
Grade-1	97.0 ± 1.05	0.90 ± 0.02	0.99 ± 0.01	0.93 ± 0.02	0.9 ± 0.02
Grade-2	88.0 ± 0.92	0.78 ± 0.03	0.93 ± 0.01	0.81 ± 0.02	0.72 ± 0.02
Grade-3	99.0 ± 0.91	1.0 ± 0.02	0.99 ± 0.01	0.95 ± 0.02	0.94 ± 0.02
Average	**93.5**	**0.89**	**0.95**	**0.89**	**0.84**

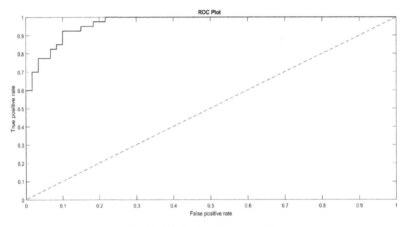

Fig. 7. ROC plot of the classifier

7   Conclusion

Computer aided automatic cancer detection and grading system for OSCC can complement the pathologist's decision and can speed up the diagnosis process. In this work, random subspace combined LDA OSCC classifier using total 83 texture and colour features has been developed, implemented and analyzed. Comparative analysis with other OSCC classifier shows that the proposed method is simple to implement and efficient with better accuracy of 93.5%, specificity 0.95 and sensitivity 0.89. Features required for this work are easy to calculate. In future work feasibility and efficiency of the same system can be tested and analyzed for other types of cancer with different staining.

Table 4. Comparison of existent methods and proposed method using LDA classifier

Ref.	Features	No. of dataset used	Performance
[9]	Morphological features of cell	Normal-9, MiD-7, SCC- 5	Accuracy: 46%
[8]	Fractal Geometry	Normal-75, MiD-52, MoD-72, SD-59, SCC-27	Sensitivity/Specificity : 63%/67%
[8]	Brownian Motion Curve	Normal-83, OSFWD-29	Accuracy: 88.89%
[7]	Statistical features	MiD-4, MoD-8, SD-10	Accuracy 89%
Current work	**Colour and texture features**	**Normal-40, WD-20, MD-30, PD-10**	**Accuracy 93.5%, Sensitivity 89%, Specificity 95%**

Note: MiD = Mild Dysplasia, MoD = Moderate Displasia, SCC = Squamous cell Carcinoma, SD = Severe Dysplasia, OSFWD = Oral Sub-mucous Fibrosys without Dysplasia

References

1. Gupta, B., Johnson, N.W.: Oral cancer: Indian pandemic. Br. Dent. J. **222**, p. 497 (2017)
2. Kerr, A.R., Shah, S.S.: Standard examination and adjunctive techniques for detection of oral premalignant and malignant lesions. J. Calif. Dent. Assoc. **41**(329–31), 334–342 (2013)
3. Koyfman, S.A., Ismaila, N., Crook, D., et al.: Management of the neck in squamous cell carcinoma of the oral cavity and oropharynx: ASCO clinical practice guideline. J. Clin. Oncol. **37**, 1753–1774 (2019). https://doi.org/10.1200/JCO.18.01921
4. Lee, S.L., Cabanero, M., Hyrcza, M., et al.: Computer-assisted image analysis of the tumor microenvironment on an oral tongue squamous cell carcinoma tissue microarray. Clin. Transl. Radiat. Oncol. **17**, 32–39 (2019). https://doi.org/10.1016/J.CTRO.2019.05.001
5. Abram, T.J., Floriano, P.N., James, R., et al.: Development of a cytology-based multivariate analytical risk index for oral cancer. Oral Oncol. **92**, 6–11 (2019). https://doi.org/10.1016/J.ORALONCOLOGY.2019.02.011
6. Ariji, Y., Fukuda, M., Kise, Y., et al.: Contrast-enhanced computed tomography image assessment of cervical lymph node metastasis in patients with oral cancer by using a deep learning system of artificial intelligence. Oral Surg. Oral Med. Oral Pathol. Oral Radiol. **127**, 458–463 (2019). https://doi.org/10.1016/J.OOOO.2018.10.002
7. Eid, R.A., Landini, G.: Quantification of the global and local complexity of the epithelial-connective tissue interface of normal, dysplastic, and neoplastic oral mucosae using digital imaging. Pathol. – Res. Pract. **199**, 475–482 (2003). https://doi.org/10.1078/0344-0338-00448
8. Muthu Rama Krishnan, M., Shah, P., Choudhary, A., et al.: Textural characterization of histopathological images for oral sub-mucous fibrosis detection. Tissue Cell (2011). https://doi.org/10.1016/j.tice.2011.06.005
9. Eid, R.A.A., Landini, G.: Oral epithelial dysplasia: can quantifiable morphological features help in the grading dilemma? In: Proceedings of the 1st ImageJ User and Developer Conference (2006)
10. Akhter, M., Rahman, Q., Hossain, S., Molla, M.: A study on histological grading of oral squamous cell carcinoma and its co-relationship with regional metastasis. J. Oral Maxillofac. Pathol. (2011). https://doi.org/10.4103/0973-029X.84485

11. Olympus CX31 Binocular Microscope - Four Objectives - Reconditioned - New York Microscope Co. https://www.microscopeinternational.com/product/olympus-cx31-binocular-microscope-four-objectives. Accessed 3 Dec 2018
12. ITU: ITU standard 709
13. Loesdau, M., Chabrier, S., Gabillon, A.: Hue and Saturation in the RGB Color Space, pp. 203–212. Springer, Cham (2014)
14. Hall-Beyer, M.: GLCM Texture: A Tutorial v. 3.0 March 2017. Arts Res. Publ. (2017) https://doi.org/10.11575/PRISM/33280
15. Kumar, R., Srivastava, R., Srivastava, S.: Detection and classification of cancer from microscopic biopsy images using clinically significant and biologically interpretable features. J. Med. Eng. **2015**, 1–14 (2015). https://doi.org/10.1155/2015/457906
16. Laws, K.I.: Rapid texture identification. In: 24th Annual Technical Symposium (1980)
17. Fisher, R.A.: The use of multiple measurements in taxonomic problems. Ann. Eugenics **7**(2), 179—188 (1936)
18. Rao, C.R.: The Utilization of Multiple Measurements in Problems of Biological Classification. J. R. Stat. Soc. Ser. B **10**, 159–203 (1948)
19. Reitsma, J.B., Glas, A.S., Rutjes, A.W.S., et al.: Bivariate analysis of sensitivity and specificity produces informative summary measures in diagnostic reviews. J. Clin. Epidemiol. (2005). https://doi.org/10.1016/j.jclinepi.2005.02.022
20. Melo, F.: Area under the ROC Curve. Encyclopedia of Systems Biology, pp. 38–39. Springer, New York, New York, NY (2013)

Automatic Annotation of Deceptive Online Reviews Using Topic Modelling

R. N. Pramukha[✉] and P. S. Venugopala

Department of CSE, NMAMIT, Nitte 574110, India
pramukhrn@outlook.com, venugopalaps@nitte.edu.in

Abstract. The increasing popularity of e-commerce websites and online review platforms has unfortunately led to the advent of review spammers. This has, in turn, led to many problems, both in business and in academia. One of the major challenges in this field is the annotation of deceptive reviews. To date, different approaches have been employed in the creation of a labelled dataset for classification tasks. Many of these works follow a general approach and do not focus on any particular property of deceptive reviews. We believe that a fine-grained approach would be more suitable for such a complex problem. This paper focuses on a single property of deceptive reviews; the out-of-context property. We first find the minimum length of review required for obtaining coherent topics. We then propose a methodology for scoring and labelling the reviews and evaluate it by training different classifiers. We obtain an F-measure of 93.64 using labelled reviews obtained through the proposed methodology.

Keywords: Deceptive review · Spam review · Topic modelling · Natural language processing · Machine learning

1 Introduction

The problem of deception in business practices is not a new one. With the advent of online review platforms and e-commerce websites, the problem has only amplified. The advancement of the internet at such an increasing rate[1] has ensured that this problem is only going to get worse. The deceptive reviews are termed as spam reviews and their authors, review spammers. Companies hire spam reviewers to not only write fake reviews to promote their business, but also to defame and hinder their competition.

Opinions shared in reviews help the customer make purchase decisions. It can also greatly influence the reputation of the product and the company. For this reason, many businesses encourage customers to write positive reviews on their product pages. Customer opinions also help the vendors in assessing the state of their business and in making business decisions. Thus, a competing business can mislead both the vendors and customers through deceptive reviews.

---

[1] https://www.cisco.com/c/en/us/solutions/collateral/service-provider/visualnetworking-index-vni/white-paper-c11-738429.html.

© Springer Nature Singapore Pte Ltd. 2020
S. Saha et al. (Eds.): MMLA 2019, CCIS 1290, pp. 41–51, 2020.
https://doi.org/10.1007/978-981-33-6463-9_4

The task of identification of deceptive reviews finds its use in academia too. There are many cases where researchers try to use reviews for tasks like sentiment analysis. If the reviews used contain deceptive reviews, the results and insights obtained by these researchers might be misleading. Therefore, the problem of review spam detection is important and essential.

Identification of deceptive reviews has already been recognized as a complex issue [1] where it is almost impossible for a human being to classify or label deceptive opinions manually. Thus, one of the main problems in the field of spam review detection is obtaining a reliable labelled dataset. Many research works have used techniques such as rules and heuristics [1–4], manual labelling by humans [4], crowdsourcing [5–10], proprietary filtering algorithms [11–13]. There is a need to identify several properties of deceptive reviews and study them in isolation. We take a fine-grained approach and concentrate on only one of the properties of deceptive reviews: the out-of-context property. A review is said to be out-of-context if it is not in context with reference to the truthful reviews. This context property is determined using topic distributions and word distributions of the reviews.

The main contributions of this paper are divided into two parts. In the first part, we determine the minimum length of review required for the extraction of good quality topics. In the second part, we propose a methodology for scoring reviews using topic modelling. This score is then used in labelling the reviews. We show the effectiveness of the proposed methodology in the task of labelling of reviews that the model has never seen before.

2 Related Work

The problem of detection of spam in digital media is not a new one. Over the past decades, spam in various media has been substantially studied. These include email spam [14, 15], SMS spam [16], web spam [17], Social media spam [18, 19].

The field of review spam is comparatively new. The early works of spam review detection were carried out by Jindal and Liu [1, 20]. They analyzed 5.8 million reviews from Amazon (amazon.com). They recognized three main categories of deceptive reviews. Type 1) Untruthful opinions or reviews which deliberately mislead the readers, Type 2) Reviews that don't comment on the product, but those which comment on the brand instead, Type 3) Reviews that are completely out of context and can be considered as non-reviews. They recognized that the process of manually labelling was difficult and tricky and they proposed to use duplicate reviews as spam reviews. Fornaciari and Poesio [3] and Hammad et al. [2] also proposed heuristics to recognize deceptive reviews and label them for classification.

Li et al. [4] employed 10 college students to annotate spam reviews based on 30 rules specified in an online study. Students worked independently and if a review gathered two out of three votes, then it was labelled as deceptive. They also found out that most of the spam reviews had relatively low helpfulness score. The problem with this approach is that manual labelling of reviews by humans can be highly subjective. The deceptive reviews are written to seem like a normal review and with an intent to mislead others. In such a scenario, manually labelling of deceptive reviews can be unreliable [21].

Yoo and Gretzel [22] gathered 42 deceptive and 40 truthful reviews. They employed students who had studied or had experience in marketing. The students were instructed to write reviews through the perspective of a hotel manager.

Labelled data were also created using AMT (Amazon Mechanical Turk) which is a crowdsourcing platform. Here, human workers were employed to write deceptive reviews instead of labelling reviews written by others. Ott et al. [5] used AMT to create 20 deceptive reviews each for 20 different hotels in Chicago. They mined 5-star reviews from TripAdvisor and used them as truthful reviews. This dataset contained only positive sentiment. In a later work, Ott et al. [6] created a negative sentiment dataset to complement the positive sentiment dataset.

Harris [7] also created a deceptive review dataset through crowdsourcing and showed that a combination of statistical assessment methods with human-based assessment provides better results than only using human-based assessors. Later, many researchers [8–10, 23] used this technique of crowdsourcing to obtain the labelled dataset. The crowdsourced reviews are considered as artificial reviews as the deceptive reviews were not obtained from real review spammers.

Mukherjee et al. [11] studied the application of existing works on real-world reviews. They found that the artificial review dataset created by Ott et al. [5] had different word distributions than real-world reviews that they obtained from TripAdvisor. They attributed this to be the reason for the high accuracy obtained by Ott et al. [5]. A real word spammer has an opportunity to analyze the existing reviews and this may influence their reviews. Human workers in platforms like AMT write reviews in isolation, hence their reviews might have completely different word distributions when compared to other reviews from the platform.

Another source of obtaining labelled data is through filtering algorithms used by sites like Yelp. These algorithms are said to be highly reliable. The major disadvantage is that the algorithms are confidential. There have been efforts to study these algorithms indirectly by using labelled datasets produced by them. Mukherjee et al. [11], Rayana and Akoglu [12] tried to do the same. Li et al. [13] created a dataset using Dianping's fake review detection algorithm. They claim that Dianping's detection algorithm has very high precision and made use of positive and unlabelled data.

Emerson et al. [24] showed that in online learning scenarios, the classification accuracy is lower in contrast to offline learning scenarios. This is important as the detection of spam reviews in real-world setting is carried out in an online learning scenario.

Due to the problem of obtaining a labelled dataset, some works employed unsupervised learning [25–27] and semi-supervised learning techniques [4, 28].

3 Topic Modelling

Topic modelling is a type of statistical modelling technique for discovering latent or abstract topics from a given set of documents. It is an unsupervised technique which assumes every document to be made up of a set of definite topics. In this paper, the context of the reviews is determined using topic modelling. The assumption is that a review is said to be in context with another review if they share common topics. This assumption is used in the annotation of spam and non-spam reviews. We use two topic modelling techniques, namely, Latent Dirichlet Allocation and Non-negative Matrix Factorization. They are described briefly in the following sections.

3.1 Latent Dirichlet Allocation

LDA [29, 30] is a Bayesian modelling technique which works with the bag-of-words representation of the documents. The key assumption is that every document is generated from a small set of topics and every topic is generated from a set of words. Before the process, the number of topics is fixed. The further inference process is described below.

Consider a corpus with M documents and K unique words.

$\theta_i \sim Dir(\alpha)$ is the topic distribution for the document i, where $i \in \{1, ..., M\}$ and $\varphi_k \sim Dir(\beta)$ is the word distribution where k, where $k \in \{1, ..., K\}$.

For each word w_i where $i \in \{1, ..., M\}$:

a) select a topic $z_i \sim Multinomial(\theta_i)$
b) select a word $w_i \sim Multinomial(\varphi_k)$

The model has 2 hyperparameters α and β. Higher the value of α, higher the topic density in documents. A higher β value ensures that topics are made up of most of the words in the vocabulary.

3.2 Non-negative Matrix Factorization

NMF is a matrix factorization technique which factorizes a document-term matrix into two matrices W and H as specified in Eq. 1.

$$M = WH \tag{1}$$

W contains the basis vectors which forms the topics discovered from the corpus and H is the coefficient matrix which contains weights for topics in each document. Both W and H contain no negative values. NMF is considered as a machine learning algorithm where W and H are calculated through optimizing an objective function. The optimization of objective function results in minimization of the reconstruction error. The W and H are iteratively derived using the update rules specified in Eq. 2.

$$W \leftarrow W \frac{MH^T}{WHH^T} H \leftarrow H \frac{W^T M}{W^T WH} \tag{2}$$

3.3 Assessing the Topic Quality

Topic quality is defined as how well the topic represents the latent semantics of a given set of documents. Most commonly used technique in the case of probabilistic modelling is the measure of log-likelihood of held-out set. Chang et al. [31] showed that perplexity (predictive likelihood) is not correlated with human judgement. They ran a topic intrusion task where they extracted 5 words from a topic and added a 6th word which was not related to that topic and shuffled these words and presented them to human judges. If the human judges could successfully pick out the intruder word, it was concluded that the words in the topic were logically connected. In other words, the topic was good at describing an idea. Roder et al. [32] carried out experiments on multiple datasets involving multiple coherence measures and concluded that topic coherence, on the other hand, is positively correlated with human judgement.

Topic coherence measures generally use an external corpus such as Wikipedia, Word-Net as a reference corpus. This works well with general documents. But, the topic modelling of reviews is a specialized task and an external reference corpus might not fully capture the word co-occurrence statistics. Mimno et al. [33] have shown that the corpus used to study the topics can also be successfully used to calculate the coherence measure. They proposed UMass coherence measure which calculates log probabilities between top word pairs. It is an asymmetrical measure which uses top words from topics to calculate the coherence score.

$$C_{UMass} = \frac{2}{N(N-1)} \sum_{i=2}^{N} \sum_{j=1}^{i-1} log \frac{P(w_i, w_j) + \epsilon}{P(w_i)} \tag{3}$$

A smoothing factor ϵ is added to the conditional probability to avoid logarithm of zero.

4 Finding the Minimum Length of Reviews

In this section, we find the minimum length of the review that contains enough information for coherent topics to be extracted from them. The dataset used was collected by [34, 35]. It contains reviews spanning from May 1996 to July 2014 from amazon.

We use two different approaches of topic modelling in this paper. The two approaches differ in Part-Of-Speech tagging phase. The first approach is the all-pos tag approach which follows the normal preprocessing phases. The second approach is the non-only approach where we only use the nouns present in the reviews. The minimum review length is calculated for both of these approaches.

For each interval of 20 units of review length, 30 topic models were trained. The 1st model consisting of 1 topic, each successive model consisting one more topic than the model preceding it, with the 30th model consisting of 30 topics. The coherence score for each interval is the arithmetic average of all the coherence scores of 30 topic models. This procedure was carried out for all-pos-tags and noun-only approaches. The resulting plots are is shown in Fig. 1 and Fig. 2.

For lower review lengths, coherence score is less as expected. In the case of all-pos tags, the convergence takes place relatively quicker than noun-only approach as reviews in noun-only approach need to be long enough to contain enough nouns to form coherent topics. After a point, the graphs seem to flatten out. No considerable increase in topic coherence is seen beyond this point which shows us that these reviews contain topics which are coherent enough that good quality topics can be extracted from them. Thus, this length can be considered as the minimum length to obtain coherent topics and only those reviews which are greater than that point is considered for further topic modelling.

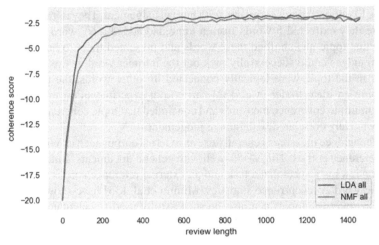

Fig. 1. Coherence score obtained for each review length interval of 20 for all-pos tags approach.

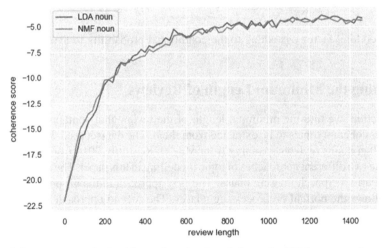

Fig. 2. Coherence score obtained for each review length interval of 20 for noun-only approach.

5 Labelling the Reviews

For this part of the research, we used reviews from 5 different categories, namely, Baby products, Digital music, Office products, Gourmet and Kitchen items and Pet supplies. A review is said to be out-of-context if it does not belong to the category in which it was originally posted. The categories which were used were themselves diverse, hence, each category themselves were found to be made up of a small subset of topics.

As stated previously, along with the usual all-pos tags approach, we also follow a second approach based on nouns present in the reviews. We hypothesize that nouns can better represent the topics belonging to categories. As the reviews were taken from a common domain (amazon.com), we expect the reviews to contain common verbs, adverbs etc. As a consequence, nouns may help distinguish between different categories. Elimination of product and people names and misspelt names is carried out before using the reviews for noun-only approach. This is to ensure that the corpus is not polluted by unwanted nouns which may give inaccurate results.

We obtain 3 types of topics from the review corpus. The first type contains particular topics which are unique to a category. The second type of topics describes two or more categories. One such topic that we obtained was about food and it described kitchen food and pet food. Consequently, it was present both in 'Gourmet and Food products' and 'Pet supplies'. The third type of topics is general topics. These are usually present in all the categories and describe a general idea about the review platform like product packaging in this case. Each category in all-pos tags approach was mostly made up of type 1 topics with few type 2 and type 3 topics. In contrast, we did not obtain any type 3 topics for noun-only approach. Categories mainly contained type 1 and some type 2 topics.

Now we describe the methodology that was used. A review dataset D is considered to be made up of i reviews and j independent categories. Each review r present in the dataset belongs to only one of the j categories.

1. Topic model T is trained for D and m topics are found. The number of topics is calculated using the coherence measure.
2. For each review in the dataset, its constituent topics are predicted using the topic model T.
3. For each category in j, the frequency of occurrence of topics for every review is aggregated. Topics with frequency > 1/m are taken as constituent topics for that category. This ensures that topics belonging to out-of-context reviews are ignored.
4. A scoring function for each review is defined as the ratio of true topics to the total number of topics present in that review.

$$S(r_{i,j}) = \frac{|T(r_i) \cap C_j|}{|T(r_i)|} \qquad (4)$$

$r_{i,j}$ represents a review i belonging to category j. Cj represents a set of topics belonging to category j. The value of this scoring function ranges from 0 to 1(with both 0 and 1 inclusive). If a review does not contain any topic from its respective category, then it gets a score of 0. If all of the topics present in the review is also present in its category, it gets a score of 1. This scoring function even takes care of reviews containing only type 3 topics by assigning it a low score.

The reviews with higher or near-perfect scores are taken as true reviews for the category and by our definition, these are the non-spam reviews. The reviews with zero or least scores are the spam reviews. These reviews do not contain topics from its own.

We trained 5 different classifiers namely, support vector machine, Bernoulli naïve Bayes, multinomial naïve Bayes, k-nearest neighbours and decision tree. Top 1000 reviews with the highest score were taken as truthful reviews and bottom 1000 reviews with the lowest score were taken as deceptive reviews. The F1 score, Precision, Recall for all the classifiers are given in Table 1 and Table 2. The results shown by the metrics only reflect how the out-of-context property performs in the classification task.

Table 1. F-measure, precision, recall obtained from different classifiers.

	SVM			B.NB			M.NB		
	P	R	F	P	R	F	P	R	F
LDA all pos	78.85	79.10	78.93	70.19	89.39	78.56	79.41	69.90	74.20
NMF all pos	90.30	86.70	88.42	77.88	66.00	71.30	79.48	80.80	80.08
LDA noun	90.45	**97.09**	**93.64**	89.59	90.90	90.20	89.96	95.00	92.38
NMF noun	**92.81**	93.49	93.13	73.34	74.80	73.98	85.17	79.00	81.92

Table 2. F-measure, precision, recall obtained from different classifiers.

	KNN			DT		
	P	R	F	P	R	F
LDA all pos	75.47	77.70	76.52	70.07	71.10	70.58
NMF all pos	75.94	84.30	79.83	79.40	74.20	76.50
LDA noun	89.37	96.29	92.68	80.80	94.50	86.96
NMF noun	79.83	92.20	85.56	81.75	80.50	82.16

Support vector machine outperforms all other classifiers in most cases. This is expected as SVM is shown to be highly effective in high dimensional vector spaces [36]. LDA noun-only approach outperforms other approaches in recall and F-measure with SVM. NMF noun-only approach outperforms all other approaches with respect to precision using SVM. We see that noun-only approach generally has better scores than all-pos tags approach.

Table 3 provides the arithmetic average of precision, recall and F-measure of all the learners for each topic modelling approach. LDA noun-only approach outperforms other approaches in all the cases. This is followed by the NMF noun-only approach.

Table 3. Average F-measure, precision, recall from different classifiers.

	Average		
	P	R	F
LDA all-pos	74.79	77.43	75.75
NMF all-pos	80.60	78.40	79.22
LDA noun	**88.03**	**94.75**	**91.17**
NMF noun	82.58	83.99	83.35

6   Effectiveness on an External Dataset

The methodology described in the previous section used the same reviews which were used to learn the topics to score the reviews. In this section, we test the effectiveness of the proposed methodology in handling reviews that the topic model has never seen before. To ensure that the reviews used for testing, truthfully represent their respective categories, we take top 1000 scoring reviews from each category and treat them as an external dataset. We learn topics using the remaining reviews. The results are given in Table 4. Accuracy represents the accuracy with which the truthful reviews are correctly classified.

Table 4. Accuracy for an external dataset.

	Accuracy
LDA all-pos	98.22
NMF all-pos	98.12
LDA noun	97.06
NMF noun	97.02

The false negatives had word distribution which the topic model had never seen before. These words never appeared in the learning dataset. This is the disadvantage of using a topic model on an external dataset. If the reviews in the external dataset are completely different from the learning dataset, the model's accuracy can be drastically affected. In the case of noun-only approach, there were brand names which were present in external dataset and which were not present in the training set. Such brand names with low frequencies can be hard to eliminate during the pre-processing stage.

7   Conclusion

In this study, we have developed a methodology to label deceptive reviews based on their out-of-context property. We followed two approaches, namely, all-pos tags and noun-only approach. We found the minimum length of reviews required for both the approaches and used topic models to score and label the reviews. Finally, we tested the effectiveness of our approach by using an external dataset which it had never encountered before.

The method we used in finding the minimum length of reviews was completely independent of finding of deceptive reviews. The results from these methods can be successfully used for any topic modelling approach involving reviews.

Even though the models were able to label the external dataset with high accuracy, it had some false negatives. We do not recommend using a topic model to label external reviews, because, depending upon the difference in word distributions in learning and external datasets, the effectiveness of this methodology can be significantly affected.

This research dealt with only one of the aspects of deceptive reviews. The out-of-context property can be used as one of the features in a large deceptive review filtering pipeline. Such a pipeline involving many fine-grained features about deceptive reviews might provide better results than a general brute-force approach. For future work, we plan on identifying additional fine-grained features based on different properties of spam reviews.

Competing Interests. The authors declare that they have no competing interests.

References

1. Jindal, N., Liu, B.: Opinion spam and analysis. In: Proceedings of the 2008 International Conference on Web Search and Data Mining, pp. 219–230 (2008)
2. Hammad, A.A., El-Halees, A.: An approach for detecting spam in Arabic opinion reviews. Int. Arab J. Inf. Technol. **12**(1), 9–16 (2015)
3. Fornaciari, T., Poesio, M.: Identifying fake Amazon reviews as learning from crowds (2015)
4. Li, F., Huang, M., Yang, Y., Zhu, X.: Learning to identify review spam. In: IJCAI International Joint Conference on Artificial Intelligence (2011)
5. Ott, M., Choi, Y., Cardie, C., Hancock, J.T.: Finding deceptive opinion spam by any stretch of the imagination, pp. 309–319, July 2011
6. Ott, M., Cardie, C., Hancock, J.T.: Negative deceptive opinion spam. In: Proceedings of NAACL-HLT 2013, pp. 497–501 (2013)
7. Harris, C.G.: Detecting deceptive opinion spam using human computation. In: Workshops at the Twenty-Sixth AAAI Conference on Artificial Intelligence (2012)
8. Ahmed, H., Traore, I., Saad, S.: Detecting opinion spams and fake news using text classification. Secur. Priv. **1**(1), e9 (2018)
9. Wilson, J., Hernández-Hall, C.: VADER: a parsimonious rule-based model for sentiment analysis of social media text. In: Eighth International AAAI Conference on Weblogs and Social Media (2014)
10. Etaiwi, W., Naymat, G.: The impact of applying different preprocessing steps on review spam detection. Procedia Comput. Sci. **113**, 273–279 (2017)
11. Mukherjee, A., Venkataraman, V., Liu, B., Glance, N.: What yelp fake review filter might be doing? In: AAAI, pp. 409–418 (2013)
12. Rayana, S., Akoglu, L.: Collective opinion spam detection: bridging review networks and metadata. In: KDD (2015)
13. Li, H., Chen, Z., Liu, B., Wei, X., Shao, J.: Spotting fake reviews via collective positive-unlabeled learning. In: Proceedings - IEEE International Conference on Data Mining, ICDM (2015)
14. Mujtaba, G., Shuib, L., Raj, R.G., Majeed, N., Al-Garadi, M.A.: Email classification research trends: review and open issues. IEEE Access **5**, 9044–9064 (2017)
15. Bhowmick, A., Hazarika, S.M.: E-mail spam filtering: a review of techniques and trends. In: Kalam, A., Das, S., Sharma, K. (eds.) Advances in Electronics, Communication and Computing. LNEE, vol. 443, pp. 583–590. Springer, Singapore (2018). https://doi.org/10. 1007/978-981-10-4765-7_61
16. Abdulhamid, S.M., et al.: A review on mobile SMS spam filtering techniques. IEEE Access **5**, 15650–15666 (2017)
17. Goh, K.L., Singh, A.K.: Comprehensive literature review on machine learning structures for web spam classification. Procedia Comput. Sci. **70**, 434–441 (2015)

18. Wu, T., Wen, S., Xiang, Y., Zhou, W.: Twitter spam detection: survey of new approaches and comparative study. Comput. Secur. **76**, 265–284 (2018)
19. Abdullah, A.O., Ali, M.A., Karabatak, M., Sengur, A.: A comparative analysis of common YouTube comment spam filtering techniques. In: 6th International Symposium on Digital Forensic and Security, ISDFS 2018 - Proceeding (2018)
20. Jindal, N., Liu, B.: Review spam detection. In: Proceedings of the 16th International Conference on World Wide Web (WWW 2007), pp. 1189–1190 (2007)
21. Bond, C.F., DePaulo, B.M.: Accuracy of deception judgments. Pers. Soc. Psychol. Rev. **10**(3), 214–234 (2006)
22. Yoo, K.-H., Gretzel, U.: Comparison of deceptive and truthful travel reviews. In: Information and Communication Technologies in Tourism 2009 (2009)
23. Li, J., Ott, M., Cardie, C., Hovy, E.: Towards a General Rule for Identifying Deceptive Opinion Spam (2015)
24. Cardoso, E.F., Silva, R.M., Almeida, T.A.: Towards automatic filtering of fake reviews. Neurocomputing **309**, 106–116 (2018)
25. Kumar, A., et al.: Spotting opinion spammers using behavioral footprints. In: KDD 2013 Proceedings of the 19th ACM SIGKDD International Conference on Knowledge Discovery and Data Mining, pp. 632–640 (2013)
26. Fei, G., Mukherjee, A., Liu, B., Hsu, M., Castellanos, M., Ghosh, R.: Exploiting burstiness in reviews for review spammer detection. In: Proceedings of the 7th International Conference on Weblogs and Social Media, ICWSM 2013 (2013)
27. Lau, R.Y.K., Liao, S.Y., Kwok, R.C.-W., Xu, K., Xia, Y., Li, Y.: Text mining and probabilistic language modeling for online review spam detection. ACM Trans. Manag. Inf. Syst. **2**(4), 1–30 (2011)
28. Fusilier, D.H., Cabrera, R.G., Montes-y-Gómez, M., Rosso, P.: Using PU-learning to detect deceptive opinion spam. In: Proceedings of the 4th Workshop on Computational Approaches to Subjectivity, Sentiment and Social Media Analysis (2013)
29. Blei, D.M., Ng, A.Y., Jordan, M.I.: Latent dirichlet allocation. J. Mach. Learn. Res. **3**, 993–1022 (2003)
30. Blei, D., Carin, L., Dunson, D.: Probabilistic topic models. IEEE Signal Process. Mag. **27**(6), 55–65 (2010)
31. Chang, J., Gerrish, S., Wang, C., Boyd-graber, J.L., Blei, D.M.: Reading tea leaves: how humans interpret topic models. In: Bengio, Y., Schuurmans, D., Lafferty, J.D., Williams, C.K.I., Culotta, A. (eds.) Advances in Neural Information Processing Systems 22, pp. 288–296. Curran Associates, Inc. (2009)
32. Röder, M., Both, A., Hinneburg, A.: Exploring the space of topic coherence measures. In: Proceedings of the Eighth ACM International Conference on Web Search and Data Mining - WSDM 2015, pp. 399–408 (2015)
33. Mimno, D., Wallach, H.M., Talley, E., Leenders, M., McCallum, A.: Optimizing semantic coherence in topic models. In: Proceedings of the Conference on Empirical Methods in Natural Language Processing, pp. 262–272 (2011)
34. He, R., McAuley, J.: Ups and downs: modeling the visual evolution of fashion trends with one-class collaborative filtering. In: Proceedings of the 25th International Conference on World Wide Web, pp. 507–517 (2016)
35. McAuley, J., Targett, C., Shi, Q., van den Hengel, A.: Image-based recommendations on styles and substitutes. In: Proceedings of the 38th International ACM SIGIR Conference on Research and Development in Information Retrieval, pp. 43–52 (2015)
36. Joachims, T.: Text categorization with support vector machines: learning with many relevant features. In: Nédellec, C., Rouveirol, C. (eds.) ECML 1998. LNCS, vol. 1398, pp. 137–142. Springer, Heidelberg (1998). https://doi.org/10.1007/BFb0026683

Machine Learning Applications

Machine Learning Technique for Analyzing the Behavior of Fish in an Aquarium

Rishabh Bhaskaran[1], Rajesh Kanna Baskaran[2(✉)], and C. Vijayalakshmi[3]

[1] Software Engineer -II, Walmartlabs, Bengaluru, India
rishabh.bhaskaran@walmartlabs.com
[2] School of Computer Science and Engineering, Vellore Institute of Technology, Chennai, India
rajeshkanna.b@vit.ac.in
[3] Department of Statistics and Applied Mathematics, Central University of Tamil Nadu, Thiruvarur, India
vijayalakshmi@cutn.ac.in

Abstract. Over the years lots of marine biologists and scientists have made efforts to study the various problem related to the ocean and its animals. One such problem faced by the fisheries is the change in the physical and mental state of the fishes when they are affected by natural calamities, external agents, change in environment, epidemic outbreak or many such factors. Under such circumstances it's probable that the fishes in the region may get affected and due to a small amount of bad fish the whole population could suffer which is a huge loss both biologically and economically. There are numerous research and studies back on the fact that fishes tend to change their behaviour in the situation of distress. The behavioural change is just in response to the distress caused to them either physically or psychologically. The motion of the fish over a period of time interval can help us learn about the behaviour it is showing. And a significant change in its pattern of motion can alarm us that there is some issue with the fish. The main idea is to identify and establish the relationship between the movements and behaviour of the fish. The video footage of a Tilapia Genus fish in a standard size aquarium set up with the help of surveillance cameras, tracked the motion of the fish in 2-D space is collected and the movement of the fish over a period of time is categorised. Based on the output the losses can be reduced.

Keywords: Epidemic · Behavior · Motion · Machine learning · Environmental change · Fourier transform · Harmonic analysis

1 Introduction

This paper mainly deals with the detailed study and categorize of fish behavior. The movements have been analyzed under water and observe changes in them over time. This also helps to identify such reactions towards environmental changes.

Objective: India is endowed with vast expanse of open inland waters in the form of canals, rivers, lakes, estuaries, reservoirs, lagoons, ponds, tanks etc. In past few years traditional aquaculture has turned into a science based commercial and economic activity

© Springer Nature Singapore Pte Ltd. 2020
S. Saha et al. (Eds.): MMLA 2019, CCIS 1290, pp. 55–65, 2020.
https://doi.org/10.1007/978-981-33-6463-9_5

involving heavy inputs and so, diseases of all kinds are known to occur on an increasingly large scale. The stressors (environment condition) which elicit morphological and physiological responses in fish fall into 4 categories as Chemical, Physical, Biological and Macro-Organisms.It would be of a great advantage if we succeed in this project and find out correlation between these stressors and the significant behavioral changes that the fishes go through. Such a system can be then implemented in aquariums and fisheries to monitor the situation in real time and also detect the potential threats by analysing the significant changes in behaviour. Fish behavior is an important factor in understanding environmental changes in water bodies [2, 7]. Marine biologists consider fish behavior to understand the upcoming changes in the environment [8]. This paper aims to create an automated system which could help determine fish behaviour and predict environment changes in water bodies. A system is proposed which makes use of Computer Vision and Machine Learning approaches to identify fish behaviour.

1.1 Setup

The first step was to setup a perfect aquarium setup (shown in Fig. 1) which makes the habitat of fish under normal circumstances. Apart from this the setup included the aerator to pump to regulate oxygen supply in the tank and the sand at the bottom to make a favourable habitat. The surveillance system included the Hikvision DS-2CE56D0T-IRP 2MP HD Indoor Night Vision Camera with Cisco IP camera switch to store the data and backup on server.

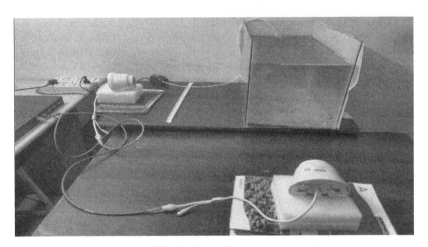

Fig. 1. Acquarium setup

2   Detection and Tracking of Fish

A computer technology related to computer image and vision processing that deals with detecting instances of semantic objects of a certain class (such as humans, fish, or house) in digital images and videos are termed as Object detection [5, 9]. In our case the subject to be tracked was our fish in the aquarium setup [4, 6]. The aim was to detect the fish in the video frames and log the coordinates of its movement over the period of time as illustrated in Fig. 2. So the values we are obtaining from the video frames are the x and y coordinates of the fish along with the time at which the fish was tracked [3, 6]. Amongst the various methods existing methods and approaches for the object detection task what we have implemented is a classification based object detection method using the Tensorflow Object Detection API [1, 10].

Fig. 2. Sample output of fish detection model

2.1 Motion Classification

Motion classification is to identify the frequently occurring motions of the fish. This will help us to understand the behaviour of the fish. We'll be able to categorise the high frequency and low frequency motion of the fish with respect to its changing position. To implement this we'll be representing the motion of the as a two dimensional time signal over its position (x, y) this will give us surface in a 3-D plot of (x, y, t).Since the tracking is done in extremely close intervals, the change in position won't be major hence we'll be considering only transition where there is a major change in position. To achieve this we'll divide the points into cluster and detect the motion between these cluster with respect to time to obtain a time-signal. Taking the Fourier transform the obtained time- signal will result in a linear combinations of sinusoid. We can easily identify the frequency of these sinusoid and this help us to determine the amount of highly-frequent motion and similarly for low- frequency motion, the amplitude of these waves will help us to identify the contribution of these signals in the final motion through the clusters.

3 Designing Approach and Details

Image recognition is important for developing most of the machine learning applications. Vision is the primary sense humans use to make decision in their everyday life. The input for the algorithm is the position of fish, this is determined using image recognition and classificaiton. To train the model we have taken snapshots of the fish from different angles and labeled them as fish.

3.1 Design Approach-Motion Classification

First step: Clustering
Since the tracking is done in extremely close intervals, the change in position won't be major hence we'll be considering only transition where there is a major change in position. To achieve this we'll divide the points into cluster and detect the motion between these cluster with respect to time to obtain a time-signal. Cluster analysis or clustering is the task of grouping a set of objects in such a way that objects in the same group (called a cluster) are more similar (in some sense or another) to each other than to those in other groups (clusters).

Elbow method:
It operates on each cluster and adds up the squares of the distance of each point in a cluster from the centroid of the cluster. That is, for each point in a cluster we take the distance between that point and the centroid of the cluster and square it. We do this for each point in the cluster and then we sum it for that cluster. Then we do this for each cluster.

When we have a "good" or tight cluster, individual distances will be small and hence the sum of squares for that cluster be small. For a "bad" or loose cluster the opposite is true. Now further, when we increase the value of K, the value of "within-cluster-sum-of-squares" will drop as we have more clusters hence smaller distances to centroids. But each successive increase in K will not give the same drop. At some point the improvement will start to level off. We call that value of K the elbow and use that as the "good" value of K.

Psuedo-Code:
For k in range(1, 10):
k = k means(data,k) #k-means clustering with k = 1..10 find_total_square_distane(data, k.centroids)Plot_total_squaredistance()

Second Step: Harmonic Wave Analysis
Harmonic wave analysis is the process of representing a time-signal as combination of sinusoids and analyzing these components with respect to their frequency, amplitude

and phase. These harmonics can used as an approximation for the signal. This enables us quick comparison, because harmonics have a standard structure.

Once we have the clusters C1.....Cn, then we can track the motion of fish through these clusters using the time parameter of <x, y, t> this will gives us a path through the clusters. Now we can think of this as a two dimensional time signal over the position of the fish. Harmonic wave analysis is procedure to split any time signal into component signals. These component signals will be linear combinations of sinusoids. To obtain this we'll be taking 2-Dimensional fourier transform on the signal.

Two Dimensional Fourier Transform

Let the signal be $y = f(x, t)$.....we have vectors of the form <x, y, t> #x,y such that x in R and y in R

Cluster.numbers $= K$

Fourier Transform of the input matrix can be calculated using:

$$X[k, l] = 1/\sqrt{MN} \sum_{n=0}^{N-1} [\sum_{m=0}^{M-1} x[m, n]e^{-j2\Pi mk/M}]e^{-j2\Pi nl/N}$$

To visualize the sinusoids, Euler's Formula:

$$e^{ix} = \cos x + i * \sin x$$

By re-grouping the coefficients of the fourier transform we can obtain the harmonics. Harmonics represent the frequency components of the wave. A higher harmonic will identify a higher frequency component signal.

Algorithm:

<x, y, t> be the coordinates representing the motion of the fish.

Fourier Transform:

Arrange <x, y, t> into $[x]_{2\times N}$ matrix where: $t \rightarrow < 0 \ldots N >$ take FFT2D of $[x]_{2\times N}$ to get $[X]_{2\times N}$

$$\begin{pmatrix} x_1 \ y_1 \\ \cdot \ \cdot \\ \cdot \ \cdot \\ \cdot \ \cdot \\ x_n \ y_n \end{pmatrix} \xrightarrow[\rightarrow]{FFT} \begin{pmatrix} X_1 \ Y_1 \\ \cdot \ \cdot \\ \cdot \ \cdot \\ \cdot \ \cdot \\ X_n \ Y_n \end{pmatrix}$$

Harmonic Analysis:

Splitting into horizontal and vertical frequency:

N^{th} harmonic will be given by:

$$\sin = imag_{part} * \sin(2\pi * n)/N$$
$$\cos = real_{part} * \cos(2\pi * n)/N$$

Where $X_0 = e^{i\theta}$, applying Euler theorem to get imag $_{part}$ and real$_{part}$.

Activity Score:

Given an input signal and a base signal

Calculate the first 7 harmonics of the input signal and base signal find the amplitude of each harmonic.

Let it be vector:

signal$_{base}$: $<A_{1b}, A_{2b},\ldots, A_{7b}>$ and signal$_{input}$: $<A_{1i}, A_{2i},\ldots, A_{7i}>$
where A_{nx} = amplitude of nth harmonic of signal x.

Then activity score of input with respect to base would be:

$$Score = \sum_{j=1}^{7} (A_{jb} - A_{ji}) * i$$

Interpretation and Inference:

Higher frequency components of the signal will be quick swimming motion made by the fish between clusters and amplitude of these waves will determine the intensity of the fish's displacement over time. Similarly low-frequency motion will describe the slow motions made by the fish in which the change in position wasn't as quick, meaning a low-frequency component of the signal.

4 Trade Offs

Selection of Appropriate Neural Network Model

As discussed earlier Tensor flow Object detection supports various models for the computation. The 5 most commonly used models are listed below. The various models provide different tradeoffs in between the two important parameters. These are the hyper parameters which provide the performance details of the models (Table 1).

Table 1. Different neural network models supported by tensor flow

Model name	Speed (ms)	COCO mAP[^1]	Outputs
ssd_mobilenet_v1_coco	30	21	Boxes
Ssd_inception_v2_coco	42	24	Boxes
Faster_rcnn_inception_v2_coco	58	28	Boxes
Faster_rcnn_resnet50_coco	89	30	Boxes

The two parameters are Speed and mAP (Mean Average Precision). As it is important that our system should be fast as well as accurate at giving results. So comparing these two parameters and fixing upon the model was an important task. One more thing that comes into picture when we talk about the neural net models is its size. We talk about portable systems we have to make sure that the model is lightweight and does not cost us much computationally. Thus the Mobile Net Model is an ideal choice for our task.

5 Conclusion

A computational system is designed and implemented which could detect, identify, study and classify the motion of the fish in the aquarium setup. The system should be able to track the aquarium and the mathematical computation over the data stream should be able to tell us about the nature of the motion of the fish in 2D space. The fish was tracked the coordinates were logged to a csv file along with timestamp, clustering was done and motion was tried to be studied using Harmonic analysis. By using the machine learning approaches and unsupervised learning techniques the behaviour of fish is analyzed along with the risk factors.

APPENDIX – Source Code

```
import numpy as np
import os
import six.moves.urllib as urllib
import sys
import tarfile
import tensorflow as tf
import zipfile

from collections import defaultdict
from io import StringIO
from matplotlib import pyplot as plt
from PIL import Image
import cv2
cap = cv2.VideoCapture('cam2.mp4')
sys.path.append("..")

from utils import label_map_util

from utils import visualization_utils as vis_util

MODEL_NAME = 'fish_detection_4'
MODEL_FILE = MODEL_NAME + '.tar.gz'
DOWNLOAD_BASE = 'http://download.tensorflow.org/models/object_detection/'
PATH_TO_CKPT = MODEL_NAME + '/frozen_inference_graph.pb'
PATH_TO_LABELS = os.path.join('data', 'fish_detection.pbtxt')
NUM_CLASSES = 1

detection_graph = tf.Graph()
with detection_graph.as_default():
  od_graph_def = tf.GraphDef()
  with tf.gfile.GFile(PATH_TO_CKPT, 'rb') as fid:
    serialized_graph = fid.read()
    od_graph_def.ParseFromString(serialized_graph)
    tf.import_graph_def(od_graph_def, name='')

label_map = label_map_util.load_labelmap(PATH_TO_LABELS)
categories      =      label_map_util.convert_label_map_to_categories(label_map,
max_num_classes=NUM_CLASSES, use_display_name=True)
category_index = label_map_util.create_category_index(categories)

def load_image_into_numpy_array(image):
  (im_width, im_height) = image.size
  return np.array(image.getdata()).reshape(
    (im_height, im_width, 3)).astype(np.uint8)
TEST_IMAGE_PATHS.
PATH_TO_TEST_IMAGES_DIR = 'test_images'
TEST_IMAGE_PATHS    =    [    os.path.join(PATH_TO_TEST_IMAGES_DIR,
```

```
'image{}.jpg'.format(i)) for i in range(1, 3) ]
    IMAGE_SIZE = (12, 8)

    with detection_graph.as_default():
    with tf.Session(graph=detection_graph) as sess:
     while True:
      time=str(cap.get(cv2.CAP_PROP_POS_MSEC))
      ret, image_np = cap.read()
      image_np_expanded = np.expand_dims(image_np, axis=0)
      image_tensor = detection_graph.get_tensor_by_name('image_tensor:0')
      boxes = detection_graph.get_tensor_by_name('detection_boxes:0')
      scores = detection_graph.get_tensor_by_name('detection_scores:0')
      classes = detection_graph.get_tensor_by_name('detection_classes:0')

      num_detections = detection_graph.get_tensor_by_name('num_detections:0')
      (boxes, scores, classes, num_detections) = sess.run(
          [boxes, scores, classes, num_detections],
          feed_dict={image_tensor: image_np_expanded})

      vis_util.visualize_boxes_and_labels_on_image_array(time,
          image_np,
          np.squeeze(boxes),
          np.squeeze(classes).astype(np.int32),
          np.squeeze(scores),
          category_index,
          use_normalized_coordinates=True,
          line_thickness=8)

      font = cv2.FONT_HERSHEY_SIMPLEX
      x = 10 #position of text
      y = 20 #position of text
      cv2.putText(image_np,time,(x,y),    font,    1,(255,255,255),2,cv2.LINE_AA)
#Draw the text

      cv2.imshow('object detection', cv2.resize(image_np, (800,600)))
      if cv2.waitKey(25) & 0xFF == ord('q'):
       cv2.destroyAllWindows()
       break
```

Function for Object Detection Visualisation

```
def visualize_boxes_and_labels_on_image_array(time,image,
                          boxes,
                          classes,
                          scores,
                          category_index,
                          instance_masks=None,
                          keypoints=None,
                          use_normalized_coordinates=False,
                          max_boxes_to_draw=20,
                          min_score_thresh=.5,
                          agnostic_mode=False,
                          line_thickness=4):

 box_to_display_str_map = collections.defaultdict(list)
 box_to_color_map = collections.defaultdict(str)
 box_to_instance_masks_map = {}
 box_to_keypoints_map = collections.defaultdict(list)
 if not max_boxes_to_draw:
  max_boxes_to_draw = boxes.shape[0]
 for i in range(min(max_boxes_to_draw, boxes.shape[0])):
  if scores is None or scores[i] > min_score_thresh:
   box = tuple(boxes[i].tolist())
   if instance_masks is not None:
    box_to_instance_masks_map[box] = instance_masks[i]
   if keypoints is not None:
    box_to_keypoints_map[box].extend(keypoints[i])
   if scores is None:
    box_to_color_map[box] = 'black'
else:
     if not agnostic_mode:
     if classes[i] in category_index.keys():
       class_name = category_index[classes[i]]['name']
      else:
       class_name = 'N/A'
      display_str = '{}: {}%'.format(
        class_name,
        int(100*scores[i]))
     else:
      display_str = 'score: {}%'.format(int(100 * scores[i]))
     box_to_display_str_map[box].append(display_str)
     if agnostic_mode:
      box_to_color_map[box] = 'DarkOrange'
     else:
      box_to_color_map[box] = STANDARD_COLORS[
        classes[i] % len(STANDARD_COLORS
```

References

1. Abadi, M., et al.: Tensorflow: large-scale machine learning on heterogeneous systems. In: Maxwell, J.C. (edn) 2015 A Treatise on Electricity and Magnetism, 3rd ed., vol. 2, pp. 68–73. Clarendon, Oxford (1892). Software available from tensorflow. Org
2. Suzuki, K., Takagi, T., Hiraishi, T.: Video analysis of fish schooling behavior in finite space using a mathematical model. Fish. Res. **60,** 3–10 (2003). K. Elissa, "Title of paper if known," unpublished
3. Siva, V., Anand, N.V., Rajesh, K.B.: Multiple live fish tracking, Yorozu, Y., Hirano, M., Oka, K., Tagawa, Y.: Electron spectroscopy studies on magneto-optical media and plastic substrate interface, IEEE Transl. J. Magn. Japan, **2**, 740–741 (1987). [Digests 9th Annual Conference Magnetics Japan, p. 301 (1982)
4. Qian, Z.-M., Cheng, X.E., Chen, Y.Q.: Automatically detect and track multiple fish swimming in shallow water with frequent occlusion. PLoS ONE **9**(9), e106506 (2014). https://doi.org/10.1371/journal.pone.0106506
5. Howard, A.G., et al.: MobileNets: efficient convolutional neural networks for mobile vision applications, Google Inc
6. Chuang, M., Hwang, J., Williams, K., Towler, R.: Tracking live fish from low-contrast and low-frame-rate stereo videos. IEEE Trans. Circuits Syst. Video Technol. **25**(1), 167–179 (2015). https://doi.org/10.1109/TCSVT.2014.2357093
7. Thida, M., Eng, H.-L., Chew, B.F.: Vision-based real-time monitoring on the behavior of fish school. In: Proceedings of the IAPR Conference on Machine Vision Applications, pp. 90–93 (2009)
8. Bong, J.C., Bong, S.B., Sam, K.C., Jung, K.O.: A simple method to quantify fish behavior by forming time-lapse images. Aquacultural Eng. **51**(2012), 15–20 (2012)
9. Zhi-Ming, Q., Shuo, H.W., Xi, E.C., Yan, Q.C.: An effective and robust method for tracking multiple fish in video image based on fish head detection
10. Build a Simple Image Recognition System with TensorFlow. Wolfgang Beyer. http://www.wolfib.com/Image-Recognition-Intro-Part-1/

Faster Convergence to N-Queens Problem Using Reinforcement Learning

Patnala Prudhvi Raj[(✉)], Preet Shah, and Pragnya Suresh

Department of Computer Science and Engineering, PES University RR Campus,
Bengaluru, India
pruthvipatnala@gmail.com, preethercules@gmail.com, pragnyasuresh@gmail.com

Abstract. Algorithmic complexity has been a constraint to solving problems efficiently. Wide use of an algorithm is dependent on its space and time complexity for large inputs. Exploiting an inherent pattern to solve a problem could be easy compared to an algorithm-based approach. Such patterns are quite necessary at cracking games with a vast number of possibilities as an algorithm-based approach would be computationally expensive and time-consuming. The N-Queens problem is one such problem with many possible configurations and realizing a solution to this is hard as the value of N increases. Reinforcement Learning has proven to be good at building an agent that can learn these hidden patterns over time to converge to a solution faster. This study shows how reinforcement learning can outperform traditional algorithms in solving the N-Queens problem.

Keywords: Reinforcement learning · Q-learning · N-Queens problem · Deep Q-learning

1 Introduction

The N-Queens problem is studied as a combinatorial constraint optimization problem which involves the placement of N queens on an $n * n$ chess board in a non-conflicting manner. A conflict is said to occur if any two queens cross each others' path. The solution requires us placing no two queens in the same column, same row or even the same diagonal. Finding a candidate solution to the N-Queens problem can be computationally expensive. Looking at the problem as choosing n places for N-Queens on an n*n board leaves us $_n^{n*n}C$ options. Restricting each queen to only a row would bring it down to n^n options. Further simplifying this, we could restrict ourselves to permutations relying on the fact that after placing a queen, it's row and column are not a part of the exploration space. This brings it down to $_n^nP = n!$ options. As n increases, the computational complexity involved in trying to find a solution increases hugely.

Though the N-Queens problem is studied merely as a mathematical recreation, it has several applications as highlighted in [1], like memory storage scheme

© Springer Nature Singapore Pte Ltd. 2020
S. Saha et al. (Eds.): MMLA 2019, CCIS 1290, pp. 66–77, 2020.
https://doi.org/10.1007/978-981-33-6463-9_6

for a conflict-free access for parallel memory systems, traffic control, modular N-queens solutions to study re-configurable meshes with buses (RMB), to find a set of deadlock-free paths, image processing in low-density parity-check codes and motion estimation. The existing approaches like Backtracking and Monte-Carlo serve well for small n values. As n increases, reaching a solution is still computationally expensive.

Expensive computation limits the wide application of algorithm-based approaches mentioned earlier. This could be mitigated by a model that kept learning inherent patterns to reach a solution faster. Such a model would be highly preferable for large values of n.

Reinforcement Learning involves an agent placed in an environment that constantly learns from its actions. The agent's action is influenced by the environment and it improves over time through the reward for its action. The present study chose to solve the N-Queens problem using reinforcement learning. Q-Learning and Deep Q-Learning were the variants chosen for this study. A study to compare the convergence rate of different algorithms (Backtracking, Monte-Carlo, Q-Learning and Deep Q-Learning) to solve the problem in minimum steps, given the same environment was also carried out.

2   Related Work

The article on Algorithms for constraint-satisfaction problems [4] shows the backtracking solution to the N-Queens problem. Counting solutions for the N-queens [11] explains the Monte-Carlo method as a probability-based algorithm to achieve the same. It has been observed that Monte-Carlo performs better than the backtracking algorithm for even values of n.

I. Rivin et al. [9] present a dynamic programming based algorithm to solve an 8×8 board problem. Here, a line is defined as the maximal co-linear set of squares on the chess board and it is said to be closed if a queen is on it. A candidate solution refers to the arrangement of n-queens in an $n \times n$ board without any conflicts. Their approach is based on the theorem that if two candidate solutions have the same set of closed lines, then the completion of one candidate is also the completion of the other. By performing the depth first search of the feasible solutions with the same set of closed lines and placing them in equivalence classes, they surpass the back-tracking algorithm which explicitly constructs all solutions to the problem.

Hu et al. [3] attempt to solve constraint optimization problems like permutations using Particle Swarm Optimization (PSO). They pick n-queens problem to demonstrate the effectiveness of their modified PSO. A particle is defined as a permutation that satisfies a constraint, a solution. With reference to the n-queens problem, a particle was defined as the permutation of row numbers of the n queens. This ensured horizontal and vertical conflicts were resolved. Hence, the objective was to eliminate the diagonal conflicts. This formed the basis for the fitness function. The fitness function was defined as the number of collisions along the diagonals of the board and the objective was to minimize the collisions along the diagonals and hence the ideal fitness value was zero.

A. Draa et al. [2] attempt to find an alternative to solve combinatorial optimization problems. They employ a quantum-inspired differential evolution algorithm to solve the n-queens problem. This algorithm is an improvisation on Differential Evolution Algorithms(DEA) and Quantum-inspired Genetic Algorithm (QGA).

The paper on Q-learning by Christopher Watkins and Peter Dayan [10] proposes a simple way for agents to learn how to optimally respond in controlled Markovian domains. It shows that Q-learning can converge to the optimum action-values with a probability of one as long as all actions are repeatedly sampled in all states and the action-values are represented discreetly.

The paper on Human-level control through deep reinforcement learning [6], gives an insight into using neural networks in conjunction with q-learning. It highlights how the DQN can out-perform the best reinforcement learning methods at various games.

The N-Queens problem was explored by Lim Soo Yeon and Son KiJun [5] as a Depth-First-Search problem. Inorder to solve the N-Queens problem by the usage of reinforcement learning, the best node from among the next available nodes had to be selected during the Depth-First-Search. Q-Learning approach was used to converge to a solution. The look-up table was simulated by assigning a value to each node. The agent would select the next best node after placing a queen and update the Q-table after every move.

Faster convergence of Q-Learning is dependent on the hyper-parameters like the learning rate α, discount factor γ and exploration rate ϵ as they decide how fast the Q-Learning model can stabilize. The paper on the acceleration of learning [8] deals with how to choose these parameters for the Q-Learning approach to stabilize quicker. The Bridge Algorithm proposed by Papavassiliou et al. [7] is robust enough to converge to an approximate global optimum for non-linear hypothesis classes as well.

The Q-learning approach [5] has been modified from a tree-based approach to a Q-table of size n^3. A Deep Q-Learning approach has also been explored by considering approaches to quickly stabilize the model by a better choice of hyper-parameters.

3   Reinforcement Learning

Reinforcement Learning is an approach modelled on how humans learn in a new environment. In this approach, an agent interacts with its environment(which can be dynamic) based on the reward it receives as it gets closer to the goal or the penalty that it incurs on moving away from the goal. Once the model learns which steps will lead to maximum reward and in turn to the goal, a policy is learnt, which is a mapping from state to action that maximizes the expected cumulative reward (value function) under that policy.

3.1   Q-Learning

Q-learning does not use the transition probability distribution that is associated with the Markov decision process and is hence a model-free reinforcement learning algorithm. Problems with stochastic transitions and rewards can be handled by it without requiring any modifications. This algorithm finds a policy that maximizes the cumulative reward which is the expected total reward it would earn on reaching the goal following the current path. This is true for any Finite Markov Decision Process (FDMP). Q-learning uses the rule:

$$Q(s_t, a_t) + = \alpha(r_{t+1} + \gamma * max_a(Q(s_{t+1}, a) - Q(s_t, a_t)))$$

Taking the maximum across all actions a which are possible at state s_t, where γ is the discounting factor which determines the weightage to be given to future rewards and α is the learning rate which determines weightage to be given to newly acquired information over old information. It makes learning independent of the starting policy π and it allows keeping this policy throughout the whole learning process (off-policy update). When Q-learning has finished, the optimal policy and the optimal value function have been found, without having to continuously update the policy during learning. Q-learning usually uses a table to store the Q-values which are updated in every iteration. At each step the algorithm chooses between exploration (finding out new ways to reach the goal) and exploitation (choosing the action that would give the highest reward referring the table). This method becomes cumbersome as the number of states and actions increase.

3.2   Deep Q-Learning

In deep Q-learning, a neural network is used to approximate the Q-value function. Since the amount of memory required to save and update the Q-table would increase as the number of states increase, the amount of time required to explore each state to create the required Q-table would be unrealistic. With the input being the state, the Q-value of all possible actions is generated as output. All the experience is stored by the user in memory as a sequence of previous actions and the corresponding rewards. The next action is determined by the maximum output of the Q-network. The loss function here is the mean squared error of the predicted Q-value and the target $Q_{value} - Q^*$. This is a regression problem. The network is going to update its gradient using back-propagation to finally converge.

4   Methods

The traditional approaches used for this comparative study were backtracking and Monte-Carlo. Two approaches under reinforcement learning have also been explored under this study. Each approach has a different initial environment to begin with. The board state is initialized to have a single queen at the bottom leftmost cell for the Q-Learning approach. The bottom row of the board is filled with queens in the Deep Q-Learning approach. The start states for both approaches can be seen in Fig. 3.

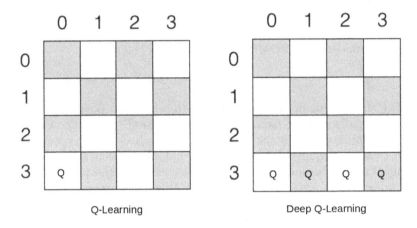

Fig. 1. Start States for n = 4

4.1   Q-Learning

In this approach, a table with all the possible states is considered. On every action, points are either awarded or reduced from a state. After the learning process, it will be possible to traverse the table to give the sequence of actions to get the solution.

Here, the q-table is designed to be a 3-dimensional array. Each cell of the board is represented by an array of width of n (array $a1$), which is in an array of width n representing which column we are in (array $a2$), which is in another array representing which row to put the queen in (array $a3$). On each turn, a queen is placed in the next column and a row is chosen for it. Each position in $a1$ corresponds to a row the previous queen can be placed in. The 3-dimensional array representation is highlighted in Fig. 2.

The q-table is initialized with zeros. The rate of exploration ϵ is set to 0.9. On every turn, first a random number between zero and one is picked and if it is lesser than ϵ, a random action (row) is chosen for the queen and if it is greater than ϵ, the q-table is referred. The value of ϵ is reduced by a factor after every iteration to reduce the rate of exploration and exploit the knowledge learnt more. After placing the queen, the board is checked to see if there are conflicts. If a conflict is found, the states that led up to this point are penalized and the board state is reset, else the procedure continues.

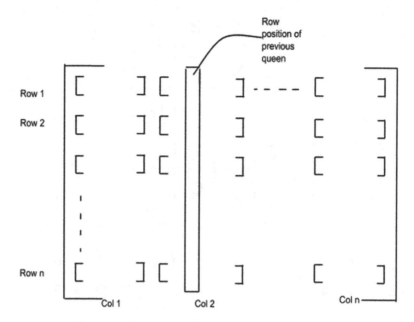

Fig. 2. Q-Table Representation

Updating Q-Table. The agent can take action using the q-table based on the sequence of actions previously taken rather than depending only on the current state. If the previous action was k, then k^{th} index in every array (row) of the next column is checked to find the max value (row index = action). This would mean that if the previous action was k, the most probable next step will be action. This is similar to having a tree with all possible states and going to the most favorable child. On selecting a cell, its value is updated as:

$$Q_{value} = Q_{value} + \alpha * (\gamma * nextmax - Q_{value})$$

where γ is the discount factor, nextmax is the max value of the next column given current action and α is the learning rate. The hyper-parameters for training can be found in Table 1.

4.2 Deep Q-Learning

Q-Learning estimates the next action based on the highest q-value entry for a particular state. This approach is quite effective for smaller values of n. Since the Q-Table representation from the Q-Learning approach has a space complexity of n^3, it is not very efficient for larger n values. Therefore, Q-Learning is effective for a limited space or environment.

Neural networks are universal function approximators. To reduce the space complexity and better approximate the q-table, a neural network can be used. This is the heart of Deep Q-Learning.

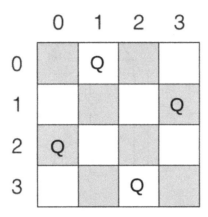

Fig. 3. Q-table update pattern for n = 4. The x mark column is decided based on the row number of the previous queen and highest value selected (red x). (Color figure online)

In the implementation, a sequential neural network model was built. The network takes the state of the chessboard as input and predicts the best action. The state is represented as a $n * n + 1$ sized vector where the last bit is used to represent which queen's action has to be predicted. The output of the network is a set of q-values from which the action as a result of the highest q-value is chosen.

The reward is calculated based on the board state as the negative of the number of conflicts. Sequence of actions and their rewards are saved in memory. These sequences are further used to train the neural network model to better approximate q-values and hence converge to optimal action sequences. The reward is optimal if an action led to the solution. Else, the target is updated as the sum of reward and argmax of prediction of the next state. The target for the current state as a result of the action leading to the next state is updated as the newly calculated target and the model is re-trained to approximate the same. This phase of training based on past sequences is called replay. The exploration rate ϵ is decreased over each episode to help achieve faster convergence.

There is a trade-off between the number of unique solutions that can be explored and the ϵ-decay rate. Greater the decay rate, faster the transition from exploring to exploiting. Uncovering all solutions would need high exploration and hence a small ϵ-decay rate would serve well. This implementation mainly focused on reaching a solution rather than finding all solutions and hence the ϵ-decay rate was high. The model ends up generalizing better than the Q-Learning approach and also saves space.

A sequential model was used to realise the architecture of the DQN Agent. The model employs a series of fully connected layers with a ReLU activation. The last layer however uses a linear activation unit. Mean squared error is used as the error measure. The neural network architecture used for the DQN agent is shown in Fig. 4 and the hyper-parameters for training can be found in Table 1.

Table 1. Hyper-parameters used for Q-Learning and Deep Q-Learning

Parameter	Q-Learning	Deep Q-Learning
Learning rate α	0.8	0.001
Discount factor γ	0.9	0.95
Exploration rate ϵ	0.9	1.0
Epsilon decay factor	0.95	0.995

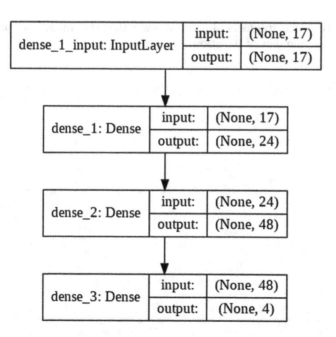

Fig. 4. DQN Agent architecture for n = 4

5   Results

The results from Q-Learning and the Deep Q-Learning approach are compared to traditional algorithms like Backtracking and Monte-Carlo. It can be seen that the reinforcement learning approaches perform better. The result is evaluated in terms of the number of iterations it takes to converge to a solution. For the reinforcement learning approaches, the iterations correspond to the 22nd episode of training (Table 2).

Table 2. Iterations to converge to a solution by various algorithms

N-Value	Back Tracking	Monte-Carlo	Q-Learning	DQN
4	26	16	4	9
5	15	16	5	7
6	171	66	6	13
7	42	43	7	13
8	876	171	8	25
9	333	383	9	55
10	975	942	10	121

The convergence graphs of Q-learning show that the initial few episodes take extremely large number of iterations.

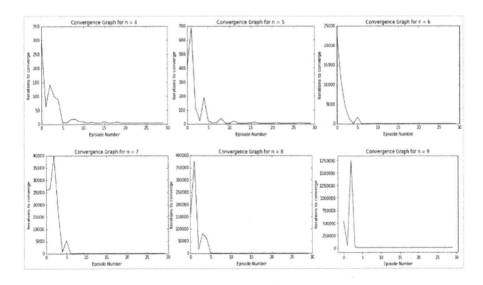

Fig. 5. Q-Learning convergence graphs for different n

The convergence graph of Deep Q-Learning below shows how the agent gets better over each episode for different n values. The number of iterations to converge to a solution is plotted against the episode number for different values of n.

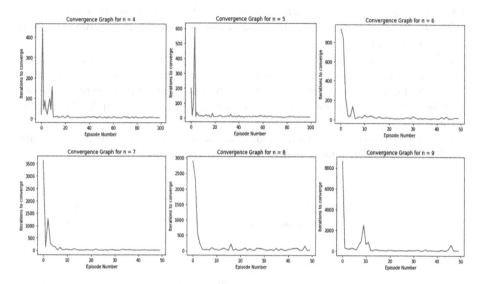

Fig. 6. Deep Q-Learning convergence graphs for different n

Convergence graphs of Deep Q-Learning resemble exponential decay as the initial episodes take a high number of iterations due to the agent actively exploring rather than exploiting. The first ten episodes of training are not strictly decaying due to exploration as seen in Fig. 7. It is during the first ten episodes that at least two different solutions to the N-Queens problem have been obtained.

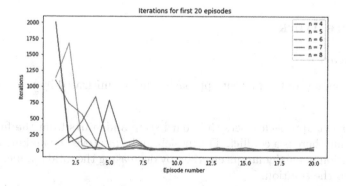

Fig. 7. Deep Q-Learning convergence during first 20 iterations

6   Conclusions

The experimental analysis using reinforcement learning approaches have led to the following conclusions:

1) Q-Learning and Deep Q-Learning ensure to converge to a solution in less than 50% of the iterations required by the traditional methods like Backtracking and Monte-Carlo.
2) Q-Learning takes a lot of iterations to converge to a solution in the initial phases of training as the entire state space is almost explored (reach a conflict and the board is reset). In contrast, Deep Q-Learning takes lesser iterations during initial stages because of the replay phase. This can be understood from the Fig. 8.
3) Q-Learning converges to a single solution and provides no scope for exploring other possible solutions in contrast to Deep Q-Learning.

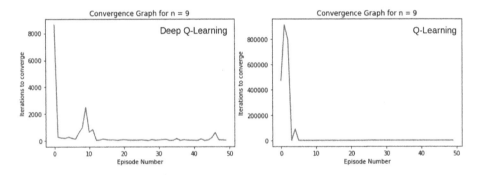

Fig. 8. Comparison of convergence for Q-Learning and Deep Q-Learning

7   Future Work

7.1   Q-Learning

The shortcomings of the present approach could be mitigated by considering the following:

1) The current approach resets the board state and starts from the first queen every time there is a conflict. The model can be improved by making it correct the position of particular queens thereby decreasing the total number of moves to get to the solution.
2) Due to exploration by trail and error, many iterations are used during the learning phase. Therefore, the number of iterations for the initial few episodes are very high and need to be reduced.

7.2   Deep Q-Learning

The proposed Deep Q-Learning method has performed really well. However, there are a few limitations in terms of the choice of queen, flexibility in movement and the need to train separately for every n. The same has been discussed here.

1) The current approach assigns turns for each queen in a round-robin fashion. The model can be improved to predict the best queen to move and further predict the action for the best queen choice.
2) The movement of a queen is restricted to a column in the current approach. This can be generalized to include horizontal and diagonal movement as well.
3) A generalized network could be designed for transfer learning so that the training time for different values of n could be reduced.

References

1. Bell, J., Stevens, B.: A survey of known results and research areas for n-queens. Discrete Math. **309**(1), 1–31 (2009)
2. Draa, A., Meshoul, S., Talbi, H., Batouche, M.: A quantum-inspired differential evolution algorithm for solving the n-queens problem. Neural Networks **1**(2) (2011)
3. Hu, X., Eberhart, R.C., Shi, Y.: Swarm intelligence for permutation optimization: a case study of n-queens problem. In: Proceedings of the 2003 IEEE Swarm Intelligence Symposium. SIS 2003 (Cat. No. 03EX706), pp. 243–246. IEEE (2003)
4. Kumar, V.: Algorithms for constraint-satisfaction problems: a survey. AI Mag. **13**(1), 32 (1992)
5. Lim, S., Son, K., Park, S., Lee, S.: The improvement of convergence rate in n-queen problem using reinforcement learning. J. Korean Inst. Intell. Syst. **15**(1), 1–5 (2005)
6. Mnih, V., et al.: Human-level control through deep reinforcement learning. Nature **518**(7540), 529 (2015)
7. Papavassiliou, V.A., Russell, S.: Convergence of reinforcement learning with general function approximators. In: IJCAI, pp. 748–757 (1999)
8. Potapov, A., Ali, M.: Convergence of reinforcement learning algorithms and acceleration of learning. Phys. Rev.E **67**(2), 026706 (2003)
9. Rivin, I., Zabih, R.: A dynamic programming solution to the n-queens problem. Inf. Process. Lett. **41**(5), 253–256 (1992)
10. Watkins, C.J.C.H., Dayan, P.: Q-learning. Mach. Learn. **8**(3), 279–292 (1992). https://doi.org/10.1007/BF00992698
11. Zhang, C., Ma, J.: Counting solutions for the n-queens and latin-square problems by monte carlo simulations. Phys. Rev. E **79**(1), 016703 (2009)

Classification of Corpus Callosum Layer in Mid-saggital MRI Images Using Machine Learning Techniques for Autism Disorder

A. Ramanathan$^{(\boxtimes)}$ and T. Christy Bobby

Department of Electronics and Communication Engineering,
Ramaiah University of Applied Sciences, Bengaluru, India
ramsaivigbala@yahoo.co.in, christy.ec.et@msruas.ac.in

Abstract. Autism is a neuro developmental disorder that affects the social interaction and communication skills of the children. It is characterized by repetitive behavior, lack of eye contact and unusual facial expressions. Corpus Callosum (CC) is the largest white matter area in the central nervous system that helps in transmission of information between both the hemispheres of brain. In autism kids, CC in the brain region shrinks and shape variations occur, making it as the region of interest with respect to diagnosis of autism disorder. Though there are many methods to segment and classify CC, there is still a need for accurate segmentation and automatic classification of CC. Since CC shares similar intensity and close proximity to other parts of the brain, segmentation of only CC region becomes challenging. To address this challenge, in the proposed work level set segmentation technique is used to segment Corpus callosum and the segmented images are validated against the ground truth using jaccard and dice index. From the segmented images geometric, texture and statistical features are extracted. Feature reduction methods such as Principal Component Analysis (PCA) and Independent Component Analysis (ICA) are incorporated for selecting the most significant set of features. Machine learning algorithms such as Support vector machine (SVM) and Extreme learning machine (ELM) are proposed to classify the image as normal and abnormal. The proposed algorithm demonstrates the classification accuracy of 97% and 96.5% using SVM and ELM respectively.

Keywords: Autism · Corpus callosum segmentation · Feature extraction · Classification

1 Introduction

Autism spectrum disorder commonly called as ASD are a group of neurological disorder that affects the social interaction and communication skills of the children [1]. Autism disorder is identified by structural changes in brain [2]. It is related to genetic mutations [3]. However exact cause of the disease is still unknown in many cases [4]. Autism is diagnosed by MRI taken from the patients. Size and shape and location of Corpus Callosum (CC) is important in identifying it from other regions of the brain. MRI is

© Springer Nature Singapore Pte Ltd. 2020
S. Saha et al. (Eds.): MMLA 2019, CCIS 1290, pp. 78–91, 2020.
https://doi.org/10.1007/978-981-33-6463-9_7

much safer technique compared to other neuro imaging techniques like skull x-rays and computed tomography (CT) [5].

Corpus callosum is the largest white matter area in the brain that connects both the right and left hemisphere [6]. CC has millions of axons connected to it [7]. It also shares close proximity with fornix, a bundle that lies below CC. Thus this CC size and shape depends on the extent of the ASD. Since CC shares a close bond with other regions of brain and has a close proximity with fornix segmenting and classifying the CC becomes extremely difficult and challenging [6]. Along with this low contrast of MRI scans and blurred CC boundaries makes the process further more challenging [5].

Though techniques like edge based active contour model, voxel intensity based segmentation, exists they suffer from disadvantages such as higher complexity, poor results due to presence of noise and less accuracy. Techniques like Active contour model do not work well if initial contour is set away from the target [6]. Although classifiers like SVM, ANN and ID3 exist, their accuracy is about 85%, 93.3% and 77% respectively [8, 9].

Thus the need for accurate segmentation and automatic classification of CC arises. To address this challenge, the proposed technique uses a combination of accurate segmentation and automatic classification. Level set method is used for accurate segmentation of CC using contours. The segmented output is validated against both jaccard and dice index to check the accuracy of segmentation. Various Geometrical, Texture statistical features are extracted from the segmented CC. Feature selection techniques like PCA and ICA are used for selecting highly significant features for better performance. The output of the feature selection techniques are individually are fed into machine learning algorithms like support vector machine and extreme learning machine which uses different kernels for classification.

The initial step is removal of skull from the brain using Matlab based on thresholding [10]. The Segmentation is carried out by the level set technique. Level set method works on both internal and external energy present within the image [11]. The segmented images are validated against the manual cropping by jaccard and dice index [12] and associated features like area, major axis, minor axis, mean, energy, entropy, homogeneity are extracted [13–16] from segmented CC.

To avoid of 'curse of dimensionality' [17] feature selection technique such as (PCA) [18] and (ICA) [19] are being used for reducing the number of features thereby taking the most significant set of features. Both PCA and ICA gives a complete new set of features which are correlated features of original variables. Classifiers such as Support vector machine (SVM) [20] and Extreme learning machine (ELM) [21] are used to classify the image as normal and abnormal. SVM is a supervised learning methodology which uses hyperplane to separates the data into different classes. To support the hyperplane kernels like Gaussian, linear and many more are used. On the other hand, ELM works on the concept that resembles the human brain. It works with the idea of hidden neurons and does not require any sort of tuning and uses kernels like sigmoid, harlim and much more.

2   Methods and Methodology

See Fig. 1.

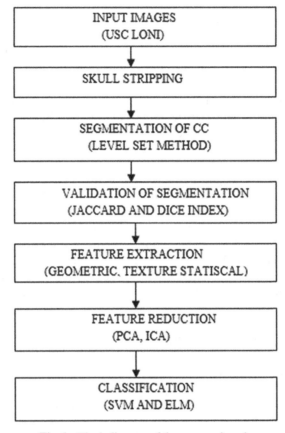

Fig. 1. Block diagram of the proposed work.

2.1   Data Set

Mid-Saggital T1 weighted brain MRI Images (N = 100) from age group of 2–40 years, collected from University of Southern California (USC)-Laboratory of Neuro Imaging (LONI) website are used for analysis. The images are in Jpeg format is of size 2048 × 2048 pixels.

2.2   Skull Stripping

The Skull is removed completely from the brain. It helps in tissue classification, brain surface reconstruction, identification of brain parts and inhomogenity correction. Skull is removed using Matlab in which the image is converted from color image to gray image and global threshold is performed. Image is then converted from gray to binary and connected components are labeled. Connected components are joined together and the unconnected part is removed.

2.3 Segmentation

Geometric active contours (GAC) use the idea of Euclidean curve shortening evolution. Contours merge and spilt depending on the object in the image. The proposed method uses GAC which makes the level set function to act as close to a signed distance function, thus completely removing the requirement of re-initialization procedure. Variational formulation comprises of an internal energy factor which adds penalty to the digression of the level set function from a signed distance function, and an external factor which controls the motion of the zero level set toward the image areas like object boundaries. The evolution of the output of level set function is the gradient flow which minimizes the total energy. It uses a larger time step and therefore speeds up the curve evolution.

The level set evolution can also be implemented by simple finite difference scheme. In level set formulation of active contour, the contours, denoted by S, are represented by the zero level set $S(t) = \{(x, y) \mid \emptyset(t, x, y) = 0\}$ of a level set function $\emptyset(t, x, y)$. Evolution of the level set function $\emptyset$ can also be written in the following general manner:

$$\partial\emptyset + \partial t + F|\nabla\emptyset| = 0 \tag{1}$$

Re-initialization is always preferred as a numerical solution in usual level set methods. The standard re-initialization method is to solve the following equation,

$$\partial\emptyset + \partial t = sign\emptyset_0(1 - |\emptyset|) \tag{2}$$

where ϕ_0 is the function to be re-initialized, and sign $(\emptyset)$ is the sign function. If ϕ_0 is not smooth or ϕ_0 is much more steeper on one side than the other, zero-level set of the final function ϕ has a chance of being moved incorrectly from original function.

Total energy functional,

$$E(\emptyset) = \mu P(\emptyset) + E_{g,\lambda,v}(\emptyset) \tag{3}$$

External energy $E_{g,\lambda,v}$ derives the zero-level set along the boundaries of the object, while the internal energy defined as $\mu P(\phi)$ penalizes the deviation of $\emptyset$ from a signed distance function during its evolution [11].

2.4 Validation

After segmentation, the segmented output is compared with region of interest (ROI) which is called as gold standard. This is created by experts using manual cropping. The segmented output and the gold standard are overlapped against each other and are checked for the overlapping coefficient or similarity index. This comparison is done by jaccard and dice similarity index techniques. This Index varies between 0 to 1 where 0 indicates no overlap and 1 denotes complete overlap. The jaccard index is given by,

$$Jaccard(A, B) = intersection(A, B)|/|union(A, B) \tag{4}$$

$$Dice(A, B) = 2^* intersection(A, B)|/|(|A| + |B|)| \tag{5}$$

|A| represents the cardinal of set A [12].

2.5 Feature Extraction

From the segmented images texture statistical features and geometrical features are extracted which are discussed briefly as follows:

Mean-Average of all the pixel values present in the image.

$$\mu = \frac{1}{NM} \sum_{i,j} p(i, j) \tag{6}$$

Skewness-It is defined as measure of symmetry and gives information about image surface.

$$\gamma_i = \left(\frac{1}{N} \sum_J^N (V_{i,j} - \mu_i)^3 \right)^{1/3} \tag{7}$$

Kurtosis-It gives information about noise and resolution measurement.

$$K_i = \left(\frac{1}{N} \sum_J^N (V_{i,j} - \mu_i)^3 \right)^{1/4} \tag{8}$$

Contrast-It measures the location variations in GLCM.

$$I = \sum \sum (x - y)^2 p(x, y) \tag{9}$$

Correlation-Gives insight about how two variables activities are associated.

$$C = \sum \sum (x - \mu x)(y - \mu y) p(x, y) / \sigma_x \sigma_y \tag{10}$$

Energy-Measures the homogeneity of the image from GLCM.

$$J = \sum_{i=1} \sum_{j=1} (p(i, j))^2 \tag{11}$$

Homogeneity-Tells about the closeness of the distribution of elements in GLCM to it's diagonal.

$$H = \sum \sum \left(\frac{p(x, y)}{1 + \{x - y\}} \right) \tag{12}$$

Area-Describes the actual number of pixels in a region. It denotes the total pixels present in the image. Thus, it tells the total area of the image.

Major Axis-Longest Diameter with respect to length in pixels. This is the line that divides the given image into 2 halves.

Minor Axis-Line that is perpendicular to major axis, this along with major axis divides the image into 2 equal halves [13–16].

2.6 Feature Selection

If more number of features is fed into the classifier as input, a phenomenon known as "Curse of Dimensionality" [17] will occur in the classification process that leads the degrading of classifier efficiency by overfiting concept. Thus, from the above derived features, the prime features are derived using dimensionality reduction techniques such as (PCA) and (ICA). PCA and ICA creates a new set of variables representing the linear combination of original variables for further analysis.

PCA is an unsupervised technique for data reduction which depicts data sets present in higher dimension to lower dimension keeping all the required linear structures steady. For dataset 'm' with respect to 'n' feature, let $k \ll n$ be the dimensionality space in which data is about to be placed and features of F are mean centered. PCA returns the top 'k' left singular vectors of F and makes the projection of newly obtained data on k dimensional subspace spread over by the columns of W_k. Then, the projector matrix in the subspace is defined as:

$$P_{W_K} = W_K W_K^T \tag{13}$$

The resulting projection obtained from the following equation, and it is reduced in all possible 'k' dimensional space.

$$\left\| F - P_{W_K} F \right\|_\xi \tag{14}$$

Where, $\xi = 2$ symbolizes spectral norm [18].

ICA is a statistical feature selection technique that eliminates the least important features and takes only independent features from the data set. The aim of ICA is to find the linear representation of the nonlinear data which are independent to each other. The n dimensional observational vectors $y = (y_1, y_2, , , , y_m)^t$ are random variables that have zero mean. Let $s = (s_1, s_2, , , , s_{d'})^t$ where d' the dimensional transform of y determining the fixed weight matrix W so that obtained variables linear transformation is,

$$s = Wy \tag{15}$$

The observed signal y can be written as independent components as

$$y = A^{-1} s \tag{16}$$

Where A is the inverse of W transform matrix [19].

2.7 Classification

In this step, the features obtained from PCA and ICA is given as input to SVM and ELM for automating the classifications of normal and abnormal CC. The performance of the classifiers for the features obtained from PCA and ICA are compared and analyzed. For training the neural network 70% of the data and for testing 30% of the data are used.

SVM is supervised algorithm that is used for classification of data. It works on the concept of finding a hyperplane that separates the similar features from data into various domains. SVM [20] is devises a computationally efficient way of separating hyperplane in a high 'n' dimensional space. The various kernels of SVM used for the analysis are as follows:

Linear kernel: It is used to separate data in to 2 classes with the help of hyperplane.

$$K(A_i, A_j) = \langle A_i.A_j \rangle \tag{17}$$

RBF kernel: This kernel maps non-linear samples into a higher dimensional space.

$$K(A_i, A_j) = e^{-}|a_i - a_j|a^2/2\sigma^2 \tag{18}$$

Polynomial Kernel: Polynomial kernel takes features combinations into account. By making $h = 1$ in the equation kernel behaves like a linear one

$$K(A_i, A_j) = (A_i, A_j + 1)^h \tag{19}$$

ELM uses hidden neurons to classify the given data. ELM consists of Multilayer of networks trained layer by layer. It is given by equation

$$\beta_i f_L(x) = \sum_{j=1}^{L} G_j(x, q_j, r_j)\beta_j \quad q_j \in R^d, r_j, \beta_j \in R \tag{20}$$

where G_j tells about the jth hidden node activation function, q_j is the input weight vector which connects the input layer to the jth hidden layer, r_j is the bias weight present at the jth hidden layer and β_j is regarded as the output weight. Various kernels like RBF, sigmoidal and polynomial kernel are used [21].

2.8 Performance Estimation

The classifier's performance is tested by sensitivity, specificity and accuracy. It is derived from the values of True Positive (TP), True Negative (TN), False Positive (FP) and False Negative (FN). TP is denoted when normal sample is classified as normal. TN comes into play when abnormal sample is classified as abnormal. Normal sample classified as abnormal is called as FP and abnormal sample classified as normal is FN [5, 22].

Sensitivity: Also called as the True positive rate (TPR) it measures the proportion of actual positives that are identified as positives. It is denoted as TPR = TP/TP + FN.

Specificity: Also called as True negative rate (TNR) it measures the proportion of actual negatives that are measured as negatives. It is denoted as TNR = TN/TN + FP.

Accuracy: Fraction of detected true samples that area actually true. (TP + TN)/(TP + TN + FP + FN.

3   Results and Discussion

Figure 2a is the MRI image of human brain in mid-saggital plane. The presence of skull and other layers of brain like cerebrum, cerebellum, and corpus callosum which is the region of interest of segmentation are seen. Figure 2b shows the contrast enhanced and

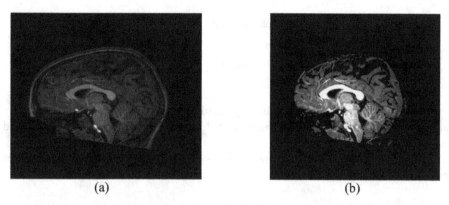

(a)   (b)

Fig. 2. a) Input image b) Skull stripped image

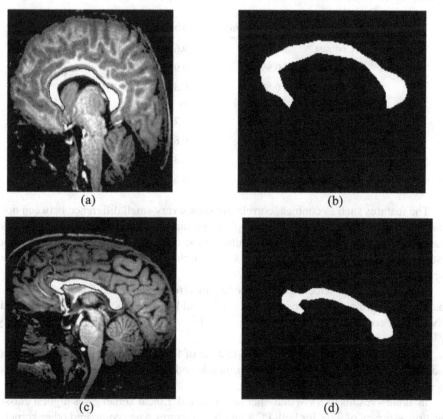

(a)   (b)

(c)   (d)

Fig. 3. a) and c): Contour present at CC. b) and d) Corresponding segmented CC alone.

skull removed image.Fig. 3a and 3c shows the contour is set at the complete layer of CC, after 20 iterations. Figure 3b and 3d shows CC alone gets segmented separately from the whole image.

Table 1. Average index of normal and autism samples

Sample	Jaccard	Dice
Normal	99.46%	99.73%
Autism	99.44%	99.72%

Table 1 shows the average values of jaccard and dice index. The higher accuracy value of these index indicates that the semented images are very close to gold standard images in the dataset.

Table 2 shows the normalized mean values of statistical and geometrical features derived from the segmented normal and autism images.

Table 2. Normalized mean values of features obtained from normal and abnormal images

Features	Normal images	Abnormal images
Contrast	0.677459	0.676456
Correlation	0.998431	0.998371
Standard deviation	0.802415	0.812949
Entropy	0.731426	0.753631
Area	0.834125	0.812949
Major axis	0.843865	0.825031
Minor axis	0.685795	0.679862

The features such as contrast, correlation shows very small difference between normal and autism brain images. However features such as area, major axis, minor axis, standard deviation and many more exhibits good demarcations between normal and autism brain images. Figure 4, 5, 6 shows some of these feature which are plotted in the form of scattergram.

PCA is applied to select best possible features from the original features. In Table 3 the first three principle components (PC1, PC2 and PC3) variance values are tabulated. These three components contribute 94% of variance and other features contribute only 6%.

Similarly ICA is also applied to original set of features and the newly transformed independent features are obtained. The principle and independent components are used as an input to SVM and ELM classifiers.

In Table 4 results of SVM classifier are tabulated. Linear kernel gives highest classification accuracy of 97% for both PCA and ICA features when compared other kernels.

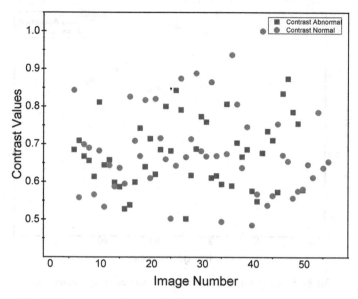

Fig. 4. Scattergram of normalized contrast values of segmented CC

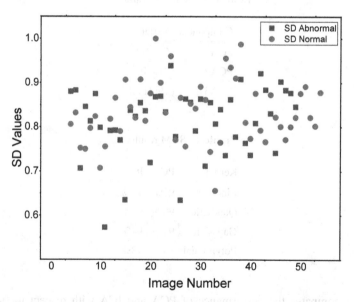

Fig. 5. Scattergram of normalized standard deviation values of segmented CC

Since highly significant features are obtained using PCA and ICA the linear classifier easily classifies the data. The classifier results shows, both PCA and ICA techniques complements each other in the SVM classification.

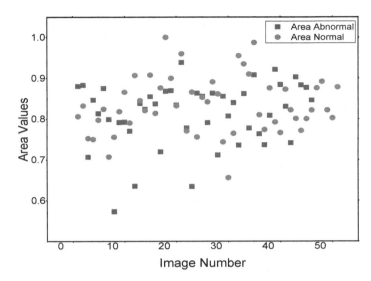

Fig. 6. Scattergram of normalized area values of segmented CC

Table 3. PCA and its variance

Component	Variance
PCA1	52.80%
PCA2	31.44%
PCA3	9.811%

Table 4. SVM results

Kernels	PCA	ICA
Linear	97%	97%
Quadratic	90%	96%
Gaussian	93%	94%
Polynomial	94%	93%

Table 5 compares the performance of PCA and ICA with respect to ELM. The training accuracy of both the feature selection techniques reaches 100% using 'Sig' kernel function. The testing accuracy also reaches the values of 96.55% in both PCA and ICA respectively showing that both techniques complement each other in the ELM classification. Hence comparing both the feature selection techniques we can conclude that both the applied feature selection techniques select highly significant set of features.

Table 5. Best obtained ELM results using both PCA and ICA techniques

Parametrs used	PCA	ICA
Kernel function	Sig function	Sig function
Number of hidden neurons	12	20
Training time	0.1404 s	0.0312 s
Training accuracy	1	1
Testing accuracy	96.55%	96.55%

SVM and ELM provides equally competitive results using both PCA and ICA indicating both the feature selection techniques gives satisfactory results.

4   Performance Estimation

Parameters such as TP, FP, TN and FN are understood with the help of confusion matrix.

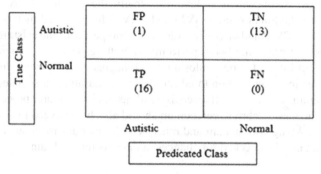

Fig. 7. Confusion matrix

Figure 7 discusses on the performance analysis of the classifiers used in the proposed method. The values of TP, TN, FP, and FN can be obtained directly from the confusion matrix. The sensitivity of the proposed system is 100% while the specificity is 92.85% respectively. Table 6 final results obtained using SVM and ELM is compared with other existing classifiers used to classify normal and abnormal patients with respect to autism [8, 9, 23].

From the table it is evident that the proposed method is much better than all the existing techniques in terms of performance parameters. Thus all the steps used in the presented work are satisfactory and quite impressive.

Table 6. Performance of existing methods in classifying autism

Existing methods	Accuracy	Sensitivity	Specificity
SVM-Linear Kernel	78.3%	76.7%	80.0%
SVM-Polynomial Kernel	83.3%	86.7%	80.0%
ANN	93.3%	96.7%	90.0%
Id3	77%	–	–
SVM	93.8%	100%	87.5%
KNN	93.8%	100%	87.5%
LDA	68.8%	62.5%	75.0%

5 Conclusions and Future Suggestions

The paper presents efficient segmentation and classification of corpus callosum in Mid-saggital MRI images for autism disorder. First, skull is removed from the brain using thresholding. Second, the corpus callosum is segmented using level set method. The segmented CC is validated using jaccard and dice index. Features associated with segmented CC are extracted and fed into feature selection techniques such as PCA and ICA to select the most significant set of features. Finally, the output of PCA and ICA are given as input to classifiers namely SVM and ELM. The proposed method reaches an accuracy of 97% in SVM and 96.5% in ELM. It also outperforms the existing techniques used in the method providing better performance with respect to sensitivity, accuracy and specificity. In future, Corpus callosum which begins to grow from 3 months from the time of pregnancy can be monitored for change in variations associated to CC and other areas of brain. This can be achieved using ultrasound with image processing. Other areas of brain such as cerebrum, cerebellum should also be analyzed to check changes in their growth. Along with autism and other neurological disorders such as epilepsy, Alzheimer's can also be checked for changes in the structure of brain.

References

1. Boger-Megiddo, I., et al.: Corpus callosum morphometrics in young children with autism spectrum disorder. J. Autism Dev. Disord. **36**(6), 733–739 (2006). https://doi.org/10.1007/s10803-006-0121-2
2. He, Q., Duan, Y., Miles, J., Takahashi, N.: A context-sensitive active contour for 2D corpus callosum segmentation. Int. J. Biomed. Imaging **2007**, 1 (2007). Article ID 24826
3. Frazier, T.W., Hardan, A.Y.: A meta-analysis of the corpus callosum in autism. Biol. Psychiatry **66**(10), 935–941 (2009)
4. Paul, L.K., Corsello, C., Kennedy, D.P., Adolphs, R.: Agenesis of the corpus callosum and autism: a comprehensive comparison. Brain **137**(6), 1813–1829 (2014)
5. Li, Y., Wang, H., Ahmed, N., Mandal, M.: Autism corpus callosum segmentation in midsagittal brain MR images. ICTACT J. Image Video Process. **8**(1), 1554 (2017)
6. Park, G., Kwak, K., Seo, S.W., Lee, J.M.: Automatic segmentation of corpus callosum in midsagittal based on Bayesian inference consisting of sparse representation error and multi-atlas voting. Front. Neurosci. **12**, 629 (2018)

7. Sakai, T., et al.: Developmental trajectory of the corpus callosum from infancy to the juvenile stage: comparative MRI between chimpanzees and humans. PLoS ONE **12**(6), e0179624 (2017)
8. Hasan, C.Z.C., Jailani, R., Tahir, N.M.: ANN and SVM classifiers in identifying autism spectrum disorder gait based on three-dimensional ground reaction forces. In: TENCON IEEE Region 10 Conference, pp. 2436–2440, IEEE, Korea (2018)
9. Bipin Nair, B.J., Ashok, G.K., Sreekumar, N.R.: Classification of autism based on feature extraction from segmented MRI image (2019)
10. Laha, M., Tripathi, P.C., Bag, S.: A skull stripping from brain MRI using adaptive iterative thresholding and mathematical morphology. In: 4th International Conference on Recent Advances in Information Technology 2018 (RAIT), pp. 1–6. IEEE, Dhanbad (2018)
11. Li, C., Xu, C., Gui, C., Fox, M.D.: Level set evolution without re-initialization: a new variational formulation. In: 2005 IEEE Computer Society Conference on Computer Vision and Pattern Recognition, vol. 1, pp. 430–436. IEEE, San Diego (2005)
12. Srinivasan, K., Nanditha, N.M.: An intelligent skull stripping algorithm for MRI image sequences using mathematical morphology. Biomed. Res. **29**(16), 3201–3206 (2018)
13. Hossain, J., Amin, M.A.: Leaf shape identification based plant biometrics. In: 13th International Conference on Computer and Information Technology, pp. 458–463. IEEE, Dhaka (2010)
14. Tarawneh, A.S., Chetverikov, D., Verma, C., Hassanat, A.B.: Stability and reduction of statistical features for image classification and retrieval: preliminary results. In: 2018 9th International Conference on Information and Communication Systems, pp. 117–121. IEEE, Jordan (2018)
15. Preethi, G., Sornagopal, V.: MRI image classification using GLCM texture features. In: 2014 International Conference on Green Computing Communication and Electrical Engineering, pp. 1–6. IEEE, Coimbatore (2014)
16. Kumar, V., Gupta, P.: Importance of statistical measures in digital image processing. Int. J. Emerg. Technol. Adv. Eng. **2**(8), 56–62 (2012)
17. Sharma, N., Saroha, K.: Study of dimension reduction methodologies in data mining. In: International Conference on Computing, Communication & Automation, pp. 133–137. IEEE, Noida (2015)
18. Boutsidis, C., Mahoney, M.W., Drineas, P.: Unsupervised feature selection for principal components analysis. In: Proceedings of the 14th ACM SIGKDD International Conference on Knowledge Discovery and Data Mining, pp. 61–69. ACM, Las Vegas (2008)
19. Cataltepe, Z., Genc, H.M., Pearson, T.: A PCA/ICA based feature selection method and its application for corn fungi detection. In: 2007 15th European Signal Processing Conference, pp. 970–974. IEEE, Poland (2007)
20. Bansal, E., Bhatia, A.: Kernel's impact on SVM classifier. In: IJARCSSE (2019)
21. Tang, J., Deng, C., Huang, G.B.: Extreme learning machine for multilayer perceptron. IEEE Trans. Neural Netw. Learn. Syst. **27**(4), 809–821 (2015)
22. Cover, G.S., Herrera, W.G., Bento, M.P., Appenzeller, S., Rittner, L.: Computational methods for corpus callosum segmentation on MRI: a systematic literature review. Comput. Methods Prog. Biomed. **154**, 25–35 (2018)
23. Oh, D.H., Kim, I.B., Kim, S.H., Ahn, D.H.: Predicting autism spectrum disorder using blood-based gene expression signatures and machine learning. Clin. Psychopharmacol. Neurosci. **15**(1), 47 (2017)

Dynamic Systems Simulation and Control Using Consecutive Recurrent Neural Networks

Srikanth Chandar[1]([✉]) and Harsha Sunder[2]

[1] Electronics and Communication Department, PES University, Bangalore, India
Srikanth.chandar@gmail.com
[2] Microspin Machine Works Pvt., Ltd., Chennai, India
harsha.skillveri@gmail.com

Abstract. In this paper, we introduce a novel architecture to connecting adaptive learning and neural networks into an arbitrary machine's control system paradigm. Two consecutive Recurrent Neural Networks (RNNs) are used together to accurately model the dynamic characteristics of electromechanical systems that include controllers, actuators and motors. The age-old method of achieving control with the use of the – Proportional, Integral and Derivative constants is well understood as a simplified method that does not capture the complexities of the inherent nonlinearities of complex control systems. In the context of controlling and simulating electromechanical systems, we propose an alternative to PID Controllers, employing a sequence of two Recurrent Neural Networks. The first RNN emulates the behavior of the controller, and the second the actuator/motor. The second RNN, when used in isolation potentially serves as an advantageous alternative to extant testing methods of electro-mechanical systems.

Keywords: RNN sequence · Electromechanical systems · Control · Simulation · PID

1 Introduction

Electromechanical systems comprise actuators, controllers and motors: in practical field work, oftentimes it is not feasible to have access to these physical systems. We propose a novel approach to their simulation and control. Currently, the field of 'Industry and Automation' lacks a definite, robust and cost-effective model to perform testing of electro-mechanical systems, and its variants across plants.

Since the 1930's, control of these systems is traditionally done through the Proportional-Integral-Derivative (PID) controllers, which is in widespread use [1, 2] (Fig. 1).

One of the most famous method of achieving control is using the Zeigler Nichols' methods. It is performed by setting the I (integral) and D (derivative) gains to zero. The "P" (proportional) gain, K_p is then increased (from zero) until it reaches the ultimate gain K_u, at which the output of the control loop has stable and consistent oscillations.

K_u and the oscillation period T_u are used to set the P, I, and D gains depending on the type of controller used.

© Springer Nature Singapore Pte Ltd. 2020
S. Saha et al. (Eds.): MMLA 2019, CCIS 1290, pp. 92–103, 2020.
https://doi.org/10.1007/978-981-33-6463-9_8

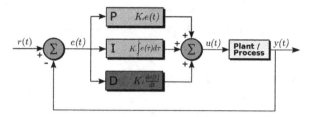

Fig. 1. PID controller equation

The correction $u(s)$ has the following transfer function relationship between error and controller output, in Laplace domain:

$$u(s) = K_p\left(1 + \frac{1}{T_i s} + T_d s\right)e(s) = K_p\left(\frac{T_d T_i s^2 + T_i s + 1}{T_i s}\right)e(s) \tag{1}$$

The disadvantages of this technique are:

- It is time consuming as a trial and error procedure must be performed.
- It forces the process into a condition of marginal stability that may lead to unstable operation or a hazardous situation due to set point changes or external disturbances.
- This method is not applicable for processes that are open loop unstable.
- Some simple processes do not have ultimate gain such as first order and second order processes without dead time.

There exist alternate methods like Tyreus-Luyben method, Damped Oscillation Method, etc. which aim to remove certain limitations of Zeigler Nichols' [3, 4]; but not entirely. The reason being that all these methods have a framework, which on a fundamental level are similar- to use a finite number of arbitrary constants which then encapsulates the entirety of a system and its behavior, regardless of the system's complexity level. The authors are of the opinion that (just) these constants are insufficient to completely describe any electro-mechanical system.

Instead of a controller being restricted to a fixed number of parameters, which then describes the system- the authors were inspired to look at the controller as a blackbox which builds a correlation between the inputs and the outputs of an electromechanical system, and thus computes the error and the suitable corrections [5] (Fig. 2).

Kwang Y. Lee [6] suggests that this blackbox can be achieved by the use of Diagonal Recurrent Neural Networks. The approach to use dynamic back propagation algorithms to these diagonal recurrent neural networks spurred the idea to use an enhanced version of the same.

It resulted in the creation of the Consecutive RNN approach, where the first RNN (RNN1) functions as an inverse of the second RNN (RNN2). RNN2 in isolation is the model which mimics the Electromechanical System (Microspin Machine [refer footnote]) to a high degree of accuracy.

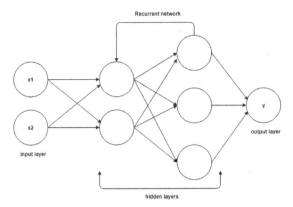

Fig. 2. RNN1 and RNN2 framework

2 Method

2.1 Machine, Model and Assumptions

The framework has two RNNs connected consecutively, and function as inverses of each other. The authors modelled an electro-mechanical system used in the textile industry by Microspin Machine Works [5] using RNNs.

The machine description is such that it takes in a time varying sequence of voltages, which is fed using Pulse Width Modulations (PWM(t)) directly proportional to the voltage at that instant. The output of this machine is corresponding time varying sequence of Revolutions Per Minute (RPM(t)).

The system in accordance to the laws of physics has an inertial lag during sudden spikes, or impulse PWMs. The prior knowledge that the authors had of the system before training the RNNs to replicate this model, is that this system has its own *flaws*- inertial lag, resonance, turbulence, lags at the start of steep function etc. The model built to replicate this machine, must have these *flaws* as well and must not be removed them in the name of efficiency (Fig. 3).

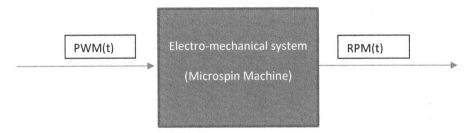

Fig. 3. Framework of electro-mechanical system

The machine has a device to capture the logged data, such as the PWM and RPM values at an instant. The logged data over a period of time had ~35,000 data points of the form (PWM, RPM). This data set comprised sinusoidal inputs, step inputs, impulse inputs and trapezoidal inputs (PWMs) - with varying slopes and peaks- and their corresponding outputs (RPMs).

An assumption made at this point is that the system is symmetric about the origin; and hence doubled the data set we had from 35,000 to 70,000 by flipping signs. This assumption is valid, since logically a negative PWM would imply a negative RPM- meaning that the motor rotates in the opposite direction.

This 70,000 data points is assumed to be a comprehensive list of details of the system, and used this as the training data to model the RNNs.

After training the RNN (RNN2 henceforth) to mimic the machine to a high degree of accuracy, it is assumed that this model is a perfect simulation of the machine, and the controller RNN (RNN1 henceforth) is trained using the original dataset, and the data from RNN2 (Fig. 4).

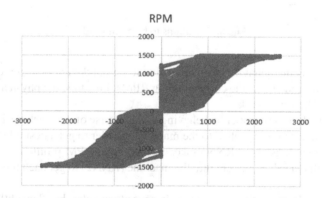

Fig. 4. Scatter plot of training data (PWM vs RPM).

2.2 Problem Formulation and Analysis

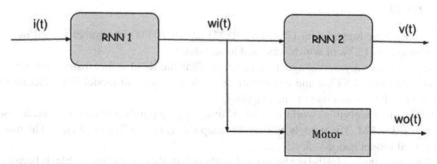

Fig. 5. Architecture of consecutive RNN approach

Where:

$i(t)$ is Target RPM profile (and input to RNN1)
$w_i(t)$ is the Predicted PWM from RNN1 and the input to machine and RNN2
$w_0(t)$ is the Actual RPM of the machine
$v(t)$ is the predicted RPM of RNN2 (model of the machine) (Figs. 5 and 6).

$$\sum_t (\omega_o(t) - v(t))^2 \qquad (1.2)$$

$$\sum_t (v(t) - i(t))^2 \qquad (1.3)$$

Fig. 6. Equations to be minimized

RNN2 models the behavior of this machine to a high degree of accuracy by taking PWMs as inputs and output the corresponding RPMs. Post this, the physical machine is removed from the framework.

RNN1 models the controller by behaving as the inverse of the machine, i.e. to predict the voltage that has to be supplied to the machine in order to get a desired RPM profile.

The whole cycle is such: RNN1 takes in the Target RPM profile and predicts the PWM profile that needs to be sent into the machine (RNN2) to get back the same Target RPM Profile.

NOTE: This framework might resemble an Autoencoder, but the subtle difference is that, in this framework- the RNN2 mimics a real life system (the electromechanical system). The output of RNN1 (or the input to RNN2) has a constraint that it should be such that it produces the Target RPM Profile, if given as inputs to the electromechanical system. This is not the case in the Autoencoder, as there is no constraint on the encoded pattern.

2.3 RNN2

- The machine logged data (PWM vs RPM), about 70,000 in number served as the training data (20% of which was used as test data).
- Given a time varying voltage, this model predicts the RPM at that instant of time.
- Its built using LSTMs and combinations of dense layers to model this function of PWM vs RPM, as a one to one mapping.
  The training algorithm works to read 'x' time varying points of PWM and predict the 'x + 1'th RPM. The models had varying step size (i.e. x = 3, x = 18 etc.). The most optimal model had x = 3.
- The reason the model did not have x = 1 (although its theoretically possible) is because 'x' cannot be too less; which would cause the model to not efficiently distinguish a data point when it has had two different histories.

- Even though RNNs account for the history through its memory feature, it was noticed that building batches of data and predicting the next in line output was more efficient. This is because the model now has a more unique data set (with a known history); so the predictions become more accurate.
- However, this batch size cannot be too large, as it would cause the system to lag – as it waits for as many intervals as the batch size, before predicting the next in line output.
- A fine balance between the two is the key.

Neural Architecture of a version of RNN2
Number of layers: 3 (2 LSTMs + 1 Dense)
Number of Neurons: 9
Number of Learnable Parameters: 122
Activation Function: Softmax
Optimizer: Adam
Loss function: Mean Square Error - between predicted output of RNN2 and true output of RNN2

$$MSE = \frac{1}{n} \sum_n \left(v_i - v_i' \right)^2$$

Where:

v_i is the ith predicted output
v_i' is the ith true output
n is the total number of points (Fig. 7).

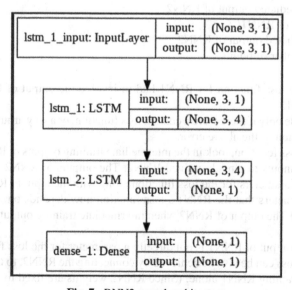

Fig. 7. RNN2 neural architecture

2.4 RNN1

- The input to RNN1 is the Target RPM Profile- at which we finally want the machine to run. We use the same training data used for training RNN2 but in an inverse fashion.
- Instead of predicting the RPM, given the PWMs; the model is trained to predict PWMs given a time varying RPM profile.
- This model is not simply an inverse of the RNN2. The method of training is completely different for the two models as they are fundamentally different in nature of data points.
- In RNN2, there was a one to one mapping of sorts between PWM and RPM, i.e. every PWM was unique and could have a corresponding RPM. The neural network was simpler, as it could uniquely relate a PWM to an RPM.
- However, with RNN1 the neural network has to be more complex, with more layers and had to be run on a greater number of epochs. This is because the RPMs were not all unique. Therefore, a prediction of an RPM, by simple logic is a one to many mappings.
- The modelling of RNN1 had to be done much differently for two reasons a) it is a one to many mapping, b) the error to minimized is:

$$MSE = \sum_t (v_i(t) - i_i(t))^2 \tag{2}$$

AND NOT:

$$MSE = \sum_t (w_i(t) - i_i(t))^2 \tag{3}$$

Where:

$v_i(t)$ is the ith predicted output of RNN2
i_i is the ith input of RNN1
w_i is the ith predicted output of RNN2
n is the total number of points.

- Therefore, the loss function for RNN1 had to involve the output of RNN2 and the input of RNN1.
- This is possible only if there were a custom loss function, or a way around that implies the minimization of the above error.
- The custom loss function, took in the intermediate training outputs of RNN1 (PWMs) and fed it as inputs to already trained RNN2. The outputs of RNN2 – RPMs, were then used as parameters for the loss function along with the input to RNN1.
- The two parameters that the RNN1's loss function now depends on, are 1) input to RNN1 and, 2) The output of RNN2 when intermediate training outputs of RNN1 are fed as input.
- Note that the output of RNN1 is not directly a parameter for the loss function.
- This custom loss can be also be achieved by connecting the RNN2, to the last layer of RNN1, and training RNN1 alone. (Since RNN2 weights are fixed to mimic the real electro-mechanical system).

Neural Architecture of a version of RNN1
Number of layers: 4 (2 LSTMs + 2 Dense)
Number of Neurons: 52
Number of Learnable Parameters: 3,553
Activation Function: Softmax
Optimizer: Adam
Custom Loss function: Mean Square Error - between output of RNN2 and input of RNN1.

$$MSE = \sum_t (v_i(t) - i_i(t))^2 \tag{4}$$

Where:
  $v_i(t)$ is the ith predicted output of RNN2
  $i_i(t)$ is the ith input of RNN1 (Figs. 8 and 9).

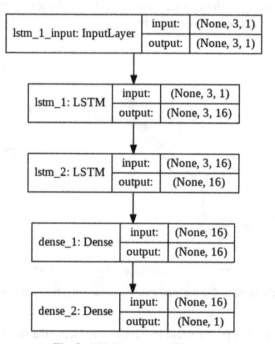

Fig. 8. RNN1 neural architecture

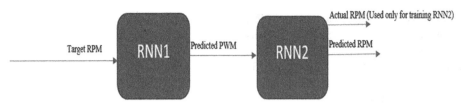

Fig. 9. Flow of consecutive RNNs

3   Results

Using Consecutive RNNs in mimicking industry equipment, particularly in the electro-mechanical systems in the textile field, has been the primary focus of this paper, and has been achieved.

The mapping between Predicted and Actual RPM (in Fig. 10) is almost spot on, with marginal error.

However, there is a significant time lag for the Predictions. The predictions are valid only after all the inputs of the step size have been fed to the RNN. There is an 18-unit time delay in the model with step size $= 18$.

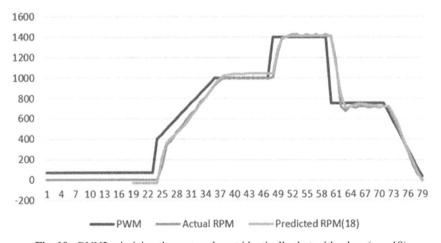

Fig. 10. RNN2 mimicing the motor almost identically; but with a lag. (x $= 18$)

Figure 11 shows that this problem is solved by reducing the step size from 18 to 3.

The following graph (Fig. 12) shows the relation between the predicted PWMs by RNN1 and the actual PWMs, when the input is a time varying RPM profile. It shows how RNN1 behaves in isolation when step size $= 1$.

Here, the step size is 1. This has been done sine the 2 RNNs are connected, which would mean that any time lag would get doubled (one for each RNN). It is noticed that this option of lag vs accuracy remains with the user, based on the situation. This lag can always be overcome by padding too.

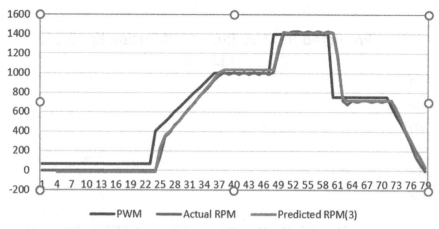

Fig. 11. RNN2 mimicing the motor almost identically; and with much lesser lag. (x = 3)

Figure 13 and Fig. 14 capture the novelty of this paper, by comparing how the given electro-mechanical system behaves when controlled by the Consecutive RNNs, as opposed to classical PID control.

It is evident that the Consecutive Recurrent Neural Networks approach works as a better controller than the classical PID controller in case of this electro-mechanical system.

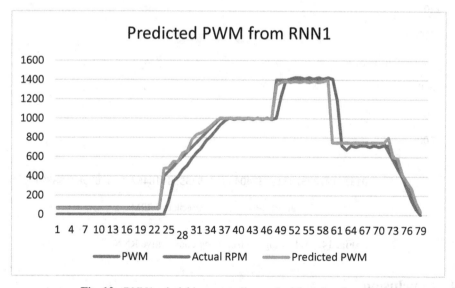

Fig. 12. RNN1 mimicking controller; and without lag. (x = 1)

The difference margins between the output and the target is visibly lesser in the case of the proposed framework in comparison to the controllers used across textile industries.

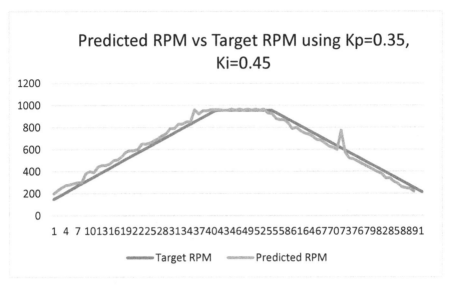

Fig. 13. PID contol using Kp, Ki constants which have been found using classical models involving trial and error.

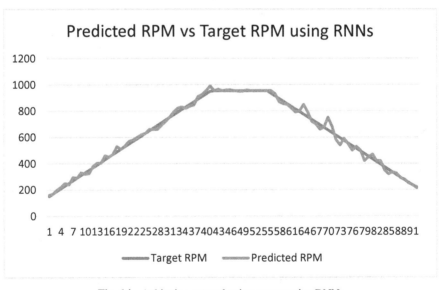

Fig. 14. Achiveing control using consecutive RNNs

4   Conclusion

In conclusion, it is evident from the graphical representation of the results that the Consecutive Recurrent Neural Network approach a) works, and b) works better than the PID controller in certain cases such as - controlling the textile motor and machinery of

Microspin Machine Works Pvt. Ltd. The authors say certain cases and do not generalize this approach to be better than PID controllers as a whole just yet, as the analysis has been with only this class of machines. However, there a lot of future scope in this domain, as this might just be the start of a new branch of Control Theory.

The past efforts that do come close to the framework proposed in this paper, do not suggest with enough conviction that this type of controller actually works on a real time dynamic electro-mechanical system [7, 8].

The other conclusions to make from this analysis is that most electromechanical systems, even the nonlinear complex ones, can be modelled using RNNs, LSTMs etc. to a good degree of accuracy. A model has been built successfully, that could save Microspin Machine Works Pvt. Ltd. and the likes, from the cost of rent, power, etc. in the factories that are set up for testing (the electromechanical systems) purposes.

Acknowledgement. As Developers at Mircospin Machine Works Pvt. Ltd., we would first thank the company and all those associated in helping to make this project a success. From running around to log the data from the Machine, to technical coding needed in training the RNNs, it's been a collective effort of many in the workplace.

We thank Mr. L Kannan, CEO of Microspin Machine Works Pvt. Ltd., for being so closely related to a project with an intern, given his stature in the company.

We also thank Dr. L Chandar (Support Vectors), and Mr. Asif Qamar (Support Vectors) who along with Mr. L Kannan, helped us in formulating the framework for this architecture. We also express our gratitude to Mr. Shritej Chavan (Student IIT-M), Mr. Pradeep Gopalakrishnan (Student IIT-M), Mr. Sumukh Nitundil (Student BITS Pilani), Ms. Shreya Vadrevu (Student PES University), Mr Chethan (Positive Integers Pvt. Ltd.), Mr. Sivaraman BV (Mircospin Machine Works Pvt. Ltd.), Prof. Rajini M (PES University) and Prof. Swetha R(PES University) for their support.

References

1. Araki, M.: PID Control. Control Systems, Robotics and Automation, vol. 2. Kyoto University, Japan
2. Bennett, S.: A brief history of automation control. IEEE Control Syst. Mag. **16**, 17–25 (2014)
3. Shahrokhi, M., Zomorrodi, A.: Comparison of PID Controller Tuning Methods'. Department of Chemical & Petroleum Engineering Sharif University of Technology
4. Atherton, D.P., Majhi, S.: Limitations of PID controllers. In: IEEE Transactions American Control Conference Brighton, UK (1999)
5. Khortabi, F.M., Khan, M.A., Potekhin, V.V.: Comparative analysis of applying deep-learning on PID Process. Automated Systems and Technologies, St. Petersberg, Russia (2017)
6. Ku, C.-C., Lee, K.Y.: Diagonal recurrent neural networks for dynamic systems control. IEEE Trans. Neural Netw. **6**, 144–156 (1995)
7. Huh, D., Todorov, E.: Real-time motor control using recurrent neural networks. In: IEEE ADPRL (2009)
8. Zhuang, M., Atherton, D.P.: Automatic tuning of optimum PID controllers. IEE PMC. Control Theoy Appl. **140**(3), 216–224 (1993)

Enhanced Hybrid Segmentation with Non Local Block and Deep Residual Networks

Yuvaram Singh$^{(\boxtimes)}$, Guda Ramachandra Kaladhara Sarma,
and Kameshwar Rao

Department of Data Analytics, HCL Technologies Limited, Noida 201304, India
yuvaramsingh94@gmail.com , grcksrgm@gmail.com, kameshjvkr@gmail.com

Abstract. The use of convolutional neural networks (CNNs) has increased in the edge devices due to its successful performance. Various such applications includes semantic segmentation which is one of the most challenging tasks due to the involvement of tremendous model size and parameters. In this paper, an enhanced hybrid segmentation with non-local block and deep residual networks is introduced for pixel level semantic segmentation. A light weight model is developed to facilitate deployment on edge devices. Skip connection is applied to fire layer in encoder and decoder block of segmentation model and non-local block is inserted in between encoder and decoder. Due to such amendments, the proposed network has optimized the number of parameters to only 2 Million whereas SegNet-Basic architecture required 5 Million. The performance is validated on the Camvid dataset and 87.5% accuracy is achieved.

Keywords: Semantic segmentation · Convolutional neural networks · SegNet · Skip connection · Non local block

1 Introduction

The progress in deep learning networks has led to lot of computer vision applications which has become part of our everyday lives. These applications include image classification, localization, object detection, segmentation and image synthesis. "Semantic segmentation is one of the high-level computer vision task that paves the way towards complete scene understanding. The importance of scene understanding as a core computer vision problem is highlighted by the fact that an increasing number of applications nourish through inferring knowledge from imagery" [3,6]. Semantic segmentation is the task of labeling the pixels of an image which belong to the same object class. Many deep learning models are developed for image segmentation in last few years; most of such networks are based on encoder-decoder approach and hence involves huge number of parameters and model size. The inference time of these models in edge devices are computationally expensive. In this work, light weight deep learning model is

S. Saha et al. (Eds.): MMLA 2019, CCIS 1290, pp. 104–110, 2020.
https://doi.org/10.1007/978-981-33-6463-9_9

presented for pixel level semantic segmentation on edge devices like NVIDIA Jetson Tx2, TI/NI board, Raspberry pi etc., The performance of proposed system is improved than the state of art methods.

This paper is organized as follows: Sect. 2 presents the proposed enhanced hybrid segmentation, Sect. 3 analyzes the performance of state-of-the-art deep learning model for pixel semantic segmentation and Sect. 4 concludes the work done.

2   Enhanced Hybrid Segmentation

In this section, an enhanced hybrid segmentation network is introduced. It consists of the non-local and deep residual networks and hence is benifited by the advantages of both. It is derived from the popular CNNs such as SegNet [1] and SqueezeNet [5] and its architecture is shown in Fig. 1.

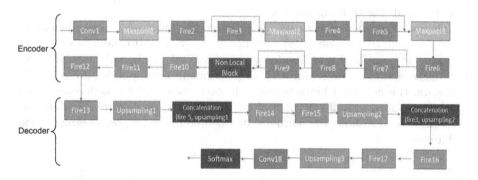

Fig. 1. Architecture of enhanced hybrid segmentation network

The proposed architecture contains encoder and decoder block. The encoder has set of one convolution, three max pooling, four fire layers with skip connection and four fire layers without skip connection. The decoder section consists of one convolution, three upsampling, eight fire layers without skip connection and two concatenation layers. The output layer with softmax activation function is used to calculate the probability of each pixel. In between encoder and decoder block, the non local neural network block is inserted to capture the dependencies within an image. The enhancement in segmentation performance is achieved due to the fire layers and non-local blocks.

The fire layer serves the purpose of reducing the number of model parameters. It has two blocks, namely squeeze and expand layer. Figure 2 presents the fire layer with and without skip connection used in this work. The squeeze layer performs 1×1 convolution to reduce the depth of output feature map. The expand layer has two convolution layers with respective filter sizes of 1×1 and 3×3, whereas the concatenation layer merges both feature maps in depth. The

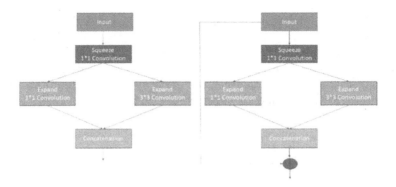

Fig. 2. Fire layer with and without skip connection

skip connection used in fire layer performs the element wise addition between input and output feature map. It helps to avoid the vanishing gradients problem and speeds up the training process.

The non-local neural network [8] captures the long range dependencies in space (images), time (Sequences) and space-time (video). It establishes the relationship between local and non-local neighborhood pixels and also computes the weighted sum of the features at all positions in the input feature maps. In this work, the non-local block is added in between encoder and decoder block which has availed to build the spatial dependencies in the feature encoding and decoding. Figure 3 shows the non-local blocks used in the proposed segmentation architecture.

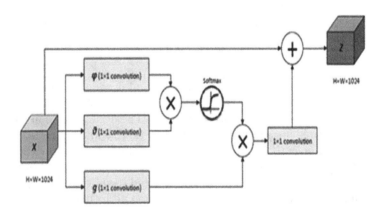

Fig. 3. Non local block

The output of generic non-local block is computed using Eq. 1 [8].

$$y_{i,j} = \frac{1}{c(x)} \sum_{\forall k,l} f(x_{i,j}, x_{k,l}) g(x_{k,l}) \tag{1}$$

where x denotes input image and y indicates the output signal with same size as that of x. (i, j) are the coordinates of position whose response is to be computed and (k, l) are the coordinates of all possible positions in the input image. $c(x)$ is used as a response normalization factor and is obtained as $\sum_{\forall k,l} f(x_{i,j}, x_{k,l})$. $f(.)$ is the similarity measurement function and $g(.)$ is a unary function which calculates the representation of input present at (k, l).

The similarity measurement function, $(f(.))$ finds the similarity between pixel $(x_{i,j})$ to all the other pixels (x_j) in an image and is given in Eq. 2.

$$f(x_{i,j}, x_{k,l}) = e^{\theta(x_{i,j})^T \phi(x_{k,l})} \tag{2}$$

Where $\theta(x_{i,j}) = W_\theta x_{i,j}$, $\phi(x_{k,l}) = W_\phi x_{k,l}$ and $g(x_{k,l}) = W_g x_{k,l}$. W_θ, W_ϕ and W_g represents the set of filter weights to perform 1×1 convolution.

Layer Name/ Type	Output Size	Filter Size / Stride	Squeeze s_{1*1}	Expand e_{1*1}	Expand e_{3*3}
Input Image	360*480*3				
Conv_1	360*480*64	3*3/1 (64)			
Maxpooling1	180*240*64	2*2/2			
Fire_2	180*240*128		16	64	64
Fire_3	180*240*128		16	64	64
Skip connection	180*240*128				
Maxpooling2	90*120*128	2*2/2			
Fire_4	90*120*256		32	128	128
Fire_5	90*120*256		32	128	128
Skip connection	90*120*256				
Maxpooling3	45*60*256	2*2/2			
Fire_6	45*60*384		48	192	192
Fire_7	45*60*384		48	192	192
Skip connection	45*60*384				
Fire_8	45*60*512		64	256	256
Fire_9	45*60*512		64	256	256
Skip connection	45*60*512				
Non-Local Block	45*60*512				
Fire_10	45*60*512		64	256	256
Fire_11	45*60*512		64	256	256
Fire_12	45*60*384		48	192	192
Fire_13	45*60*384		48	192	192
UpSampling1	90*120*384				
Concatenation (UpSampling1,Fire_5)	90*120*640				
Fire_14	90*120*256		64	128	128
Fire_15	90*120*256		64	128	128
UpSampling2	180*240*256				
Concatenation (UpSampling2,Fire_3)	180*240*384				
Fire_16	180*240*128		32	64	64
Fire_17	180*240*128		16	64	64
UpSampling3	360*480*128				
Conv_18	360*480*12	1*1/1(12)			

Fig. 4. Layer wise detailed architecture of proposed network

There are various forms of functions $f(.)$ and $g(.)$ in non-local networks. This includes Gaussian, embedded Gaussian, dot product and concatenation. The embedding Gaussian function computation is similar to the softmax function and hence preferred in this work. In this case, the non-local output is found as per mentioned in Eq. 3.

$$y_{i,j} = softmax(X^T W_\theta^T W_\phi X)g(X) \tag{3}$$

The output of the non-local block is computed by combining the original signal with non-local operation result. It is presented in Eq. 4.

$$z_{i,j} = W_z y_{i,j} + x_{i,j} \tag{4}$$

The detailed architecture of proposed network is shown in Fig. 4; the layer wise operation type, output feature map size and filter size are specified.

3   Results and Discussion

The proposed segmentation algorithm is implemented on NVIDIA Quadro P1000 (4 GB) GPU card, 32 GB RAM, intel core i7 processor in windows 10 OS. The python libraries viz., Keras, Tensorflow, Numpy, Matplotlib and OpenCV are used for coding. The CamVid [2] dataset is considered for validation of proposed method. It contains 701 samples images out of which 367, 101 and 233 samples are used for training, validation and testing, respectively. It has 12 classes of objects such as training Sky, Building, Pole, Road, Pavement, Tree, Sign Symbol, Fence, Car, Pedestrian, Bicyclist and Unlabelled. The image size is $360 \times 480 \times 3$. Adadelta optimization technique and cross entropy loss function are used for training. The training is done for 100 epochs with the batch size of 2. Table 1 indicates the performance of proposed light weight deep learning model. The enhanced hybrid segmentation network obtains the accuracy of 87.5% compared to SegNet-Basic which is 86.11%. The number of parameters is reduced to 2M from 5M. The results are also compared with existing methods such as Non local block [9], fire layers with skip connection [4,5].

Table 1. Performance evaluation of proposed method

Sr. no	Methods	Number of parameters	Accuracy
1	Segnet-Basic [1]	5467500	86.11%
2	SegNet-SqueezeNet [7]	1607564	76.42%
3	Segnet-SqueezeNet with residual connection	1476492	82.35%
4	Segnet-squeeze with skip connection and feature map concatenation	1564604	87.08%
5	**Enhanced hybrid segmentation (proposed)**	**2088892**	**87.54%**

Figure 5 gives the visual segmentation results for CamVid dataset. First row represents the original images of CamVid dataset and the segmented label of the original image is given in second row. Third row presents the reference SegNet-Basic segmented result and fourth row is the SegNet and SqueezeNet hybrid segmentation architecture result. Fifth row represent hybrid segmentation with residual connection segmented result and sixth row represents the hybrid segmentation with residual connection and skip connection. Last row shows the enhanced hybrid segmentation (non-local block based segnet squeezenet deep residual networks segmented) results.

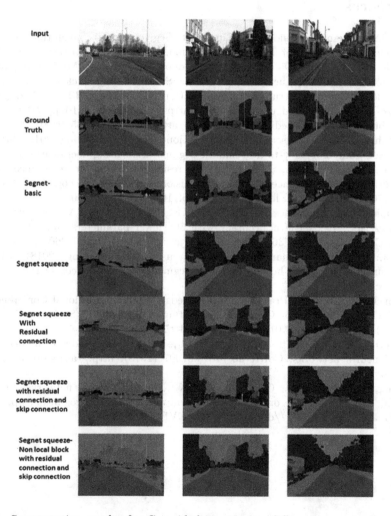

Fig. 5. Segmentation results for Camvid dataset using different segmentation techniques

4 Conclusion

In this paper we have proposed an enhanced hybrid segmentation network which consists of non-local block and deep residual neural network for pixel level semantic segmentation. The proposed segmentation architecture obtains the accuracy of 87.5% and reduced the number of parameters to 2.08M. In future, we are planning to implement different variants of skip connection and feature map concatenation for encoder and decoder block in segmentation network.

References

1. Badrinarayanan, V., Kendall, A., Cipolla, R.: SegNet: a deep convolutional encoder-decoder architecture for image segmentation. IEEE Trans. Pattern Anal. Mach. Intell. **39**(12), 2481–2495 (2017). https://doi.org/10.1109/TPAMI.2016.2644615
2. Brostow, G.J., Fauqueur, J., Cipolla, R.: Semantic object classes in video: a high-definition ground truth database. Pattern Recogn. Lett. **30**(2), 88–97 (2009). Video-based object and event analysis. https://doi.org/10.1016/j.patrec.2008.04.005. http://www.sciencedirect.com/science/article/pii/S0167865508001220
3. Gandhi, A.: How to do semantic segmentation using deep learning (2018). https://nanonets.com/blog/how-to-do-semantic-segmentation-using-deep-learning/
4. He, K., Zhang, X., Ren, S., Sun, J.: Deep residual learning for image recognition. In: 2016 IEEE Conference on Computer Vision and Pattern Recognition (CVPR), pp. 770–778, June 2016. https://doi.org/10.1109/CVPR.2016.90
5. Iandola, F.N., Moskewicz, M.W., Ashraf, K., Han, S., Dally, W.J., Keutzer, K.: SqueezeNet: AlexNet-level accuracy with 50x fewer parameters and <1mb model size. CoRR abs/1602.07360 (2016). http://arxiv.org/abs/1602.07360
6. Li, J.: How to do semantic segmentation using deep learning (2018). https://medium.com/nanonets/how-to-do-image-segmentation-using-deep-learning-c673cc5862ef
7. Bapi, R.S., Rao, K.S., Prasad, M.V.N.K. (eds.): First International Conference on Artificial Intelligence and Cognitive Computing. AISC, vol. 815. Springer, Singapore (2019). https://doi.org/10.1007/978-981-13-1580-0
8. Shokri, M., Harati, A., Taba, K.: Salient object detection in video using deep non-local neural networks. CoRR abs/1810.07097 (2018). http://arxiv.org/abs/1810.07097
9. Wang, X., Girshick, R., Gupta, A., He, K.: Non-local neural networks. In: 2018 IEEE/CVF Conference on Computer Vision and Pattern Recognition, pp. 7794–7803, June 2018. https://doi.org/10.1109/CVPR.2018.00813

Astronomy and AstroInformatics

Chaotic Quantum Behaved Particle Swarm Optimization for Multiobjective Optimization in Habitability Studies

Arun John[1][✉] and Anish Murthy[2]

[1] Computer Science, University of Alabama in Huntsville, Huntsville, USA
arunjoh@gmail.com
[2] Epsilon, Bangalore, India

Abstract. In this paper, based on the Quantum-behaved Particle Swarm Optimization algorithm in [1–3], we evolve the algorithm to optimize a multiobjective optimization problem, namely the Cobb Douglas Habitability function which is based on "CES production functions" in Economics. We also propose some changes to the Quantum-behaved Particle Swarm Optimization algorithm to mitigate the problem of the algorithm prematurely converging and show the results of the proposed changes to the Quantum-behaved Particle Swarm Optimization.

Keywords: Habitability score · Metaheuristic optimization · LDQPSO

1 Introduction

Quantum-behaved Particle Swarm Optimization (QPSO) algorithm, proposed by Jun Sun, is an evolution of the Particle Swarm Optimization originally proposed by Kennedy and Ebenhart in 1995.

Particle Swarm Optimization (PSO) is an evolutionary optimization technique, which is modelled to simulate the evolution of knowledge of a social organism, in which the individuals, which represent the candidate solutions of a particular problem, fly or move through a multidimensional space to find an optima or sub optima. These particles in the solution space of the problem are characterised by a "position" and a "velocity" in the multidimensional space and reorient their position to a goal (defined by their fitness function) in every iteration of their search algorithm. The particles in a local neighborhood share memories of their best positions and use this local knowledge as well as their own best solution to adjust their velocities.

A major draw of PSO is its simplicity and parallelizable nature. Saha et al. [4] describe PSO as "a distributed method that requires simple mathematical operators and short segments of code, making it an optimal solution where computational re-sources are at a premium. Its implementation is highly parallelizable and scales with the dimensionality of the search space. The standard

PSO algorithm does not deal with constraints but, through variations in initializing and updating particles, constraints are straightforward to represent and adhere to."

Quantum-behaved Particle Swarm Optimization (QPSO) is a quantum model of the original PSO where the state (similar to the "position" and "velocity" in PSO) of a particle is depicted by a wave-function, given by "$\psi(\vec{x}, t)$" [2], instead of a position and velocity. The dynamic behavior of the particle is different from the particle in PSO as the "position" and the "velocity" of the particles cannot be determined simultaneously. Only the probability of a particle appearing in a particular location "$\vec{X}$" can be determined from the probability density function "$|\psi(\vec{x}, t)|^2$" [2]. A delta potential well is employed to constrain the quantum particles and prevent explosion. Since the search space and the solution space are different, a state transformation from the quantum state to classical state called "collapse" is applied.

The proposed changes to the QPSO algorithm are related to the initialization of the particles as well as the position update rule for the algorithm. A chaotic initialization of the particles is done using the Lorenz attractor, which is a set of chaotic solutions for the Lorenz equation. The particle position update rule is changed to something similar to a Levy Flight mechanism, which is exhibited by animals when searching for food in an area.

The multi-objective problem that the algorithm will be fine tuned to optimize is the bi-objective Cobb Douglas Habitability function, which is used to generate the Cobb Douglas Habitability Score for exoplanets. The score is composed of two parts, namely the interior score and the surface score of the particular planet.

2   Cobb Douglas Habitability Function

The general motivation for using Cobb-Douglas production function is because of its interesting properties as described in [5]. It is a function that models the response of an output parameter on varying its inputs. According to [5], "the function is concave when the sum of elasticities is not greater than one, ensuring that an optimum exists which maximizes the function inside a feasible region defined by the constraints on elasticities". It was first originally introduced to model the growth of American economy during 1899–1922. In the case of exoplanetary habitability, the proposed metric models how the habitability score Y changes on varying input planetary parameters. This is achieved by allowing the coefficients of elasticity to be adjusted via an optimization algorithm. It has already been established that the proposed habitability metric consists of two components: surface score and interior score. The final CDHS, defined in equation, is equal to the convex combination of Y_i and Y_s. The weights ω_i and ω_s defines the importance of interior score and surface score in determining the final CDHS, respectively. Here, ω_i and ω_s sum up to 1. The Cobb-Douglas Habitability production function can be formally written as (from [5])

$$Y = R^\alpha . D^\beta . V_e^\delta . T_s^\gamma \tag{1}$$

where R, D, V_e and T_s is the radius, density, escape velocity and surface temperature respectively. α, β, δ and γ are coefficients of elasticity and $0 < \alpha, \beta, \gamma, \delta < 1$.

The Cobb Douglas Habitability score is estimated by breaking it up into the interior score ($CDHS_i$) and the surface score ($CDHS_s$) and maximizing the following production functions.

$$Y_i = CDHS_i = R^\alpha.D^\beta \tag{2a}$$

$$Y_s = CDHS_s = V_e^\gamma.T_s^\delta \tag{2b}$$

Equations (2a) and (2b) are convex under either Constant Returns to Scale (CRS), when $\alpha + \beta = 1$ and $\gamma + \delta = 1$, or Decreasing Returns to Scale (DRS), when $\alpha + \beta < 1$ and $\gamma + \delta < 1$. The final Cobb Douglas Habitability Score is the convex combination of the individual interior and surface scores, given by,

$$Y = \omega_i.Y_i + \omega_s.Y_s \tag{3}$$

+

3 Quantum-Behaved Particle Swarm Optimization

Quantum-behaved Particle Swarm Optimization is an improved version of the biologically inspired metaheuristic algorithm known as Particle Swarm Optimization, which is used to find the global minima of a function. In PSO, particles move around and converge towards the globally optimal solution while losing kinetic energy as they approach the solution, similar to how a particle would behave in a potential field of attraction at the optimal point. QPSO builds upon this by making use of quantum potential fields, and introducing the particles as quantum particles represented by their waveforms. Making use of a potential model, we can simulate the similar behaviour of particles being attracted to the centre of the quantum potential field. In most cases, the Delta Potential Well model is used for QPSO as it provides faster convergence, and this paper employs the same as introduced in [2].

3.1 Proposed Changes to the QPSO Algorithm

Chaotic Initialization. Chaos theory is a part of mathematics that looks at very sensitive systems where a very small change can make the system behave drastically differently. It deals with nonlinear events which are impossible to predict or control, like weather, turbulence, stock market etc. It is popularly known by the butterfly effect, in which the flapping of a butterfly's wings lead to a chain of events that could lead to a hurricane somewhere else. It may take a long time to become a hurricane, but the connection still exists. Since the weather is a very sensitive system, the flapping of the wings at that point in space-time or a different time would have drastically different effects. This is the a simple example of a small change in the initial conditions leading to drastic changes over time.

Edward Lorenz, who was a meteorologist-mathematician, also known as the founder of modern Chaos Theory made a weather model which involved 12 differential equations and exhibited chaotic behavior. In his effort to find chaotic systems in simpler set of equations, he was led to the phenomenon of rolling fluid convection and came up with the following equations.

$$\frac{dx}{dt} = \sigma(y - x) \tag{4a}$$

$$\frac{dy}{dt} = x(\rho - z) - y \tag{4b}$$

$$\frac{dz}{dt} = xy - \beta z \tag{4c}$$

When the parameters of the system, σ, ρ and β are 10, 28 and $\frac{8}{3}$ respectively, the system described by Equations (4a), (4b) and (4c) displays chaotic behavior.

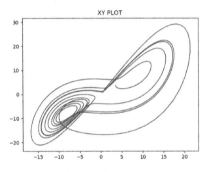

Fig. 1. Projection of the Lorenz Chaos system on the XY plane

From Figs. 1, 2 and 3 we can see that the Lorenz system of equations is a "strange attractor". Wikipedia describes "strange attractors" to be very sensitive to initial conditions, hence, any two arbitrarily close initial points on the attractor, after any numbers of iterations, will lead to points that are arbitrarily far apart (within the limits of the attractor), and after any number of iterations will lead to points that are arbitrarily close together. Thus, a system with a chaotic attractor such as the Lorenz strange attractor is locally unstable yet globally stable: i.e. once some sequences have entered the attractor, nearby points diverge from one another but never depart from the attractor.

This behavior of the Lorenz system can be used to initialize particles in the Quantum-behaved Particle Swarm Optimization algorithm. A similar approach was followed in [6], where the CPSO algorithm used the Henon map and Tent Map as the chaotic initializations of the particles. In a similar way, the Lorenz system of equations is used to create a map which initializes all the particles in the modified QPSO algorithm. Since the Lorenz system is restricted to three dimensions, multiple Lorenz systems with different initializations are used for

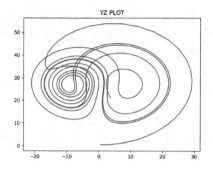

Fig. 2. Projection of the Lorenz Chaos system on the YZ plane

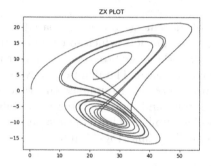

Fig. 3. Projection of the Lorenz Chaos system on the ZX plane

particles with higher dimensions. The dimensions are then scaled and mapped to the particles using the limits of the Lorenz system. The main objective behind the chaotic initialization is the importance of the initial positions of the particles as they can help resolve premature convergence which hinders the algorithm from finding the global minima of a given objective function.

Levy Flight. Levy Flight is a random walk in which the step-lengths have a probability distribution that is heavy-tailed. When defined as a walk in a space of dimension greater than one, the steps made are in isotropic random directions. As described by wikipedia, "Levy flight stems from the mathematics related to chaos theory and is useful in stochastic measurement and simulations for random or pseudo-random natural phenomena. Examples include earthquake data analysis, financial mathematics, cryptography, signals analysis as well as many applications in astronomy, biology, and physics. For general distributions of the step-size, satisfying the power-like condition, the distance from the origin of the random walk tends, after a large number of steps, to a stable distribution due to the generalized central limit theorem, enabling many processes to be modeled using Lévy flights. The probability densities for particles undergoing a Levy flight can be modeled using a generalized version of the Fokker–Planck equation, which is usually used to model Brownian motion. The equation requires the use of fractional derivatives. For jump lengths which have a symmetric probability distribution, the equation takes a simple form in terms of the Riesz fractional derivative. In one dimension, the equation reads as,

$$\frac{\delta\phi(x,t)}{\delta t} = -\frac{\delta}{\delta x}f(x,t)\phi(x,t) + \gamma\frac{\delta^\alpha\phi(x,t)}{\delta|x|^\alpha} \tag{5}$$

where γ is a constant akin to the diffusion constant, α is the stability parameter and f(x, t) is the potential. The Riesz derivative can be understood in terms of its Fourier Transform."

$$F\left[\frac{\delta^\alpha\phi(x,t)}{\delta|x|^\alpha}\right] = k^\alpha F_k\left[\phi(x,t)\right] \tag{6}$$

This naturalistic form of movement can be compared to organisms wandering away from regions of over-saturation, which in case of optimization problems is highly beneficial in allowing the model to explore a larger region in the solution space before complete convergence. The main objective in using Levy Flight in the QPSO model is that it is possible to simulate the wandering of particles away from global or known optima and improve the search abilities of the model for problems which have a high number of local optima, leading to greater frequency of convergence to the global optima (Figs. 4 and 5).

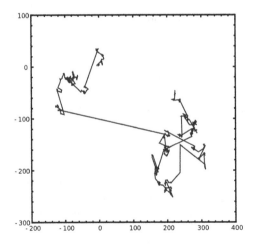

Fig. 4. An example of 1000 steps of a Lévy flight in two dimensions

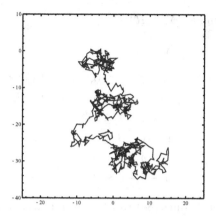

Fig. 5. An example of 1000 steps of an approximation to a Brownian motion type of Lévy flight in two dimensions

4   Representing the Problem

A Constrained Optimization problem can be represented as,

$$\underset{x}{\text{minimize}} \quad f(x)$$
$$\text{subject to} \quad g_k(x) \leq 0, \quad k = 0 \ldots N - 1,$$
$$h_l(x) = 0, \quad l = 1 \ldots r$$

Saha et al. [2] make use of a method by Ray and Liew [7], who describe a way to represent non strict inequality constraints when optimizing using a particle swarm as "strict inequalities and equality constraints are to be converted to non strict equalities before representing them in the problem. Introducing an error threshold ϵ converts the strict inequalities of the form $g'_k(x) < 0$ to non strict inequalities of the form $g_k(x) = g'_k(x) + \epsilon \leq 0$. A tolerance τ is used to convert the equality constraints into a pair of inequalities,"

$$g_{(q+l)}(x) = h_l(x) - \tau \leq 0, \quad l = 1 \ldots r,$$
$$g_{(q+r+l)}(x) = -h_l(x) - \tau \leq 0, \quad l = 1 \ldots r. \tag{7}$$

In the above way, r equality constraints become 2r equality constraints, raising the total number of constraints for to equation to $s = q + 2r$. In these, For each potential solution p_i, c_i represents the constraint vector where, $c_{ik} = \max\{g_k(p_i), 0\}$, $k = 1 \ldots s$. When $c_{ik} = 0$, $\forall k = 1 \ldots s$, the solution lies within the feasible region. When $c_{ik} > 0$, the potential solution p_i violates the kth constraint.

4.1   Representing CDH Score Estimation

Similar to the way the CDH Score is represented in [4], the CDH Score estimation under CRS is represented as,

$$\underset{\alpha,\beta,\gamma,\delta}{\text{minimize}} \quad Y_i = R^\alpha . D^\beta \tag{8a}$$

$$Y_s = V_e^\gamma . T_s^\delta \tag{8b}$$

$$\text{subject to} \quad -\phi + \epsilon \leq 0, \quad \forall \phi \in \{\alpha, \beta, \gamma, \delta\}, \tag{8c}$$

$$\phi - 1 + \epsilon \leq 0, \quad \forall \phi \in \{\alpha, \beta, \gamma, \delta\}, \tag{8d}$$

$$\alpha + \beta - 1 \leq 0, \tag{8e}$$

$$1 - \alpha - \beta \leq 0, \tag{8f}$$

$$\gamma + \delta - 1 \leq 0, \tag{8g}$$

$$1 - \gamma - \delta \leq 0. \tag{8h}$$

Under DRS, the constraints (8e) to (8h) are replaced with,

$$\alpha + \beta + \epsilon - 1 \leq 0 \tag{9a}$$

$$\gamma + \delta + \epsilon - 1 \leq 0 \tag{9b}$$

5   Experiments and Results

5.1   Testing the Proposed Algorithm

The proposed changes in the algorithm are first tested on a series of test functions given by

- Rosenbrock Function: $f(x) = (1 - x)^2 + 100(y - x^2)^2$
- Mishra Bird Function: $f(x, y) = e^{(1-\cos y)^2} \sin x + e^{(1-\sin x)^2} \cos y + (x - y)^2$
- Ackley Function: $f(x.y) = -20e^{0.2^2 \sqrt{(x^2+y^2)}} + e^{0.5+\cos 2\pi x + \cos 2\pi y} + e + 20$
- Levi function: $f(x, y) = \sin^2(3\pi x) + (x - 1)^2(1 + \sin^2(3\pi y)) + (y - 1)^2(1 + \sin^2(2\pi y))$

The algorithm was named LQPSO and another variant of it was named as the LDQPSO. The LDQSP algorithm added a Levy Flight decay, which decayed the effect of the Levy flight as the number of iterations increased. For all the algorithms, the total number of particles were 1000 and the LQPSO and the

Table 1. Avg iterations to convergence

Test functions	Algorithms			
	PSO	QPSO	LQPSO	LDQPSO
Ackley	194	32.6	33.834	45.2
Levi		130.934	46.734	48.734	49.734
Rosen		134.767	42.8	42.934	46.8
Mishra Bird		152.734	77.534	65.634	60.534

Algorithm 1. LDQPSO minimization

1: **procedure** MINIMIZE(fun, x) ▷ Minimizing function fun using initialised particles x0
2:    **repeat**
3:        $pbest \leftarrow x$
4:        $gbest = $ getBest($fun, pbest$)   ▷ Best solution in $pbest$ for fun
5:        **for** $i \leftarrow 1$ to *populationsize* M **do**
6:            **if** fun(x_i) $<$ fun($pbest_i$) **then**
7:                $pbest_i = x_i$
8:            **end if**
9:            $u = $ rand($0, 1$)
10:           $f_1 = $ rand($0, 1$),$f_2 = $ rand($0, 1$)
11:           $P = (f_1 * gbest + f_2 * pbest)/(f_1 + f_2)$
12:           find *mbest*
13:           **for** $d \leftarrow 1$ to *dimension* D **do**
14:               $l = DecayedLevyWalkFactor()$
15:               $update = mbest_d * l * \ln(1/u_d)$
16:               **if** random($0, 1$)> 0.5 **then**
17:                   $position_d = P_d - update$
18:               **else**
19:                   $position_d = P_d + update$
20:               **end if**
21:           **end for**
22:           $gbest = $ getBest($fun, pbest$)
23:       **end for**
24:   **until** termination criteria is met
25: **end procedure**

LDQPSO algorithms were initialized using Lorenz Chaos Map. The Lorenz Map is a similar one as used in [6]. Each of the algorithms were tested a total of 30 times to get their average iterations and to calculate their success rate. The results are presented in Table 2.

The results presented in Table 1 were generated using the Psopy library which was created as part of the paper in [4]. All the algorithms had a 100% success rate on all the test functions except for the Mishra Bird function. PSO had the lowest success rate of 83%, with QPSO having a better success rate of 90%, with the LQPSO and the LDQPSO both having a success rate of 93%. This shows that the proposed algorithm does better at avoiding the local optima compared to the original PSO as well as the revised QPSO algorithm.

5.2 Testing the Algorithm on the CD-HPF

After seeing the results of the LDQPSO algorithm on the test functions, the LDQPSO algorithm was used to optimize the Cobb Douglas Habitability function using a modified version of the jMetalPy framework [8]. A subset of the original PHL-EC Dataset [9] was used, specifically the exoplanets belonging to the TRAPPIST-1 system. The Pareto front plots were generated using 25 particles just as in [4] (Figs. 6 and 7).

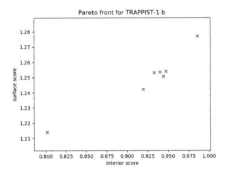

Fig. 6. Trappist-1b under CRS conditions

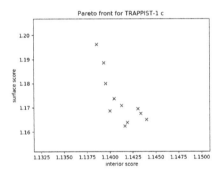

Fig. 7. Trappist-1c under CRS conditions

After comparing the results of the algorithm with the catalog mentioned in [4,5], we noticed that in most cases the points on the plot showed better scores than those mentioned in the catalog. In the original paper [5] however, the plots showed that most of the points were close together and a proper front could be seen. We think that this is not visible in the Pareto front plots generated by us as the conflict between the objective functions (for calculating CDHS) might be minimal and not visible in these plots as the convergence may be faster. Table 2 shows the average number of iterations over 30 runs that the LDQPSO algorithm took to converge for both Constant Returns to Scale as well as Decreasing Returns to Scale. A point to note is that unlike the offset in [4], there is no offset in LDQPSO, due to which it may look misleading that the LDQPSO takes more iterations than PSO, but that is not the case as there is no offset used. We can clearly see that LDQPSO algorithm requires lesser number of iterations to converge (Figs. 8, 9, 10, 11, 12, 13, 14, 15, 16 and 17) (Table 3).

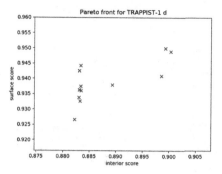

Fig. 8. Trappist-1d under CRS conditions

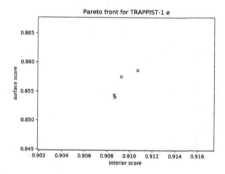

Fig. 9. Trappist-1e under CRS conditions

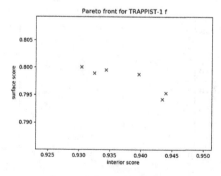

Fig. 10. Trappist-1f under CRS conditions

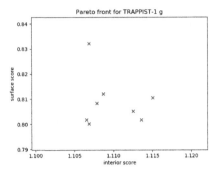

Fig. 11. Trappist-1g under CRS conditions

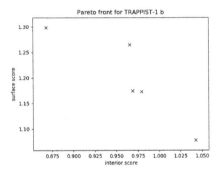

Fig. 12. Trappist-1b under DRS conditions

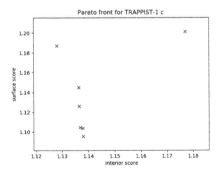

Fig. 13. Trappist-1c under DRS conditions

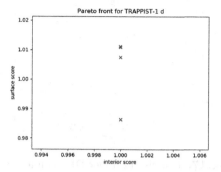

Fig. 14. Trappist-1d under DRS conditions

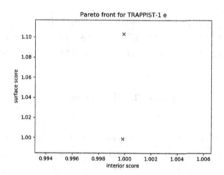

Fig. 15. Trappist-1e under DRS conditions

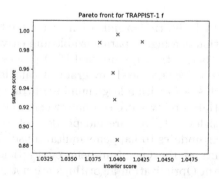

Fig. 16. Trappist-1f under DRS conditions

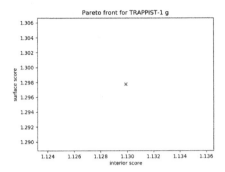

Fig. 17. Trappist-1g under DRS conditions

Table 2. Avg iterations to convergence

Returns to scale	Planets					
	Trappist-1 B	Trappist 1 C	Trappist 1 D	Trappist 1 E	Trappist 1 F	Trappist 1 G
CRS	28.3	26.067	29.567	28.034	28.3	26.734
DRS	52.3	65.034	37.034	32.834	73.767	90.3

Table 3. CDHS scores

Conditions	Planets					
	Trappist-1 B	Trappist 1 C	Trappist 1 D	Trappist 1 E	Trappist 1 G	Trappist 1 H
CRS (0.5i + 0.5s)	1.1929	1.114	0.7926	0.887	1.065	0.7382
DRS (0.5i + 0.5s)	1.1887	1.1717	1.009	0.9976	1.0795	0.9645
CRS (0.6i + 0.4s)	1.1612	1.132	0.8005	0.888	1.0730	0.7452
DRS (0.6i + 0.4s)	1.1619	1.1675	1.0073	0.9977	1.0803	0.9695

6   Conclusion

Quantum-behaved Particle Swarm Optimization as a variant of the original Particle Swarm Optimization is a highly parallelizable and easy to implement algorithm, which performs better than the original PSO proposed by Kennedy and Ebenhart in [10]. Since it does not need any gradient calculation, it can work in high dimensional search spaces with a large number of constraints, which is useful in cases such as a Habitability score estimate where many input parameters can be used. The particles in QPSO are independent of each other in a single iteration, allowing their updates to happen simultaneously and asynchronously.

Although the results of the Quantum-behaved Particle Swarm Optimization and Particle Swarm Optimization algorithms are not as accurate as direct methods, the scaling of the algorithms when the number of input parameters increases allows it to be more feasible than traditional optimization methods as it can handle the higher number of constraints.

The main aim of this manuscript is to compare the performance of the Quantum-behaved Particle Swarm Optimization with the Particle Swarm Optimization while proposing some changes to the model itself. These changes are influenced by Chaos Theory and the movement of animals foraging for food in an area. As we observed from our experiments, the modified QPSO algorithm performed better than the PSO in terms of performance and it's ability to avoid getting stuck in a local optima.

7   APPENDIX

7.1   Gradient Simulation of QPSO

The QPSO system functions by initializing a set of particles each with a pseudo-random wave function. The position of the particles as obtained from these wave functions at each iteration describe each particle's solution at that instance, which are feasible at initialization. However, the position of the particle is updated on every iteration of the process which might put the particle on an infeasible solution. Now, we simulate a particle well for each particle such that each particle well has a centre at point p, which is related to the wave function of the particle by,

$$d^2\psi/dy^2 = 2m/h[E + \gamma\delta(y)]\psi \tag{10}$$

Hence, at each iteration, the algorithm stores a set of feasible solutions L represented by the position of each particle, pi as well as the globally optimal solution gbest represented by pg. At the start of the process, the algorithm initializes L to the initial positions of the particles and gbest to the best solution in L. At each iteration, QPSO calculates the position of each particle at that instance using their p value and the gbest value pg by simulating the delta potential well with a characteristic length l determined by the gbest as,

$$L = 1/\beta = \hbar/(m\gamma) \tag{11}$$

The algorithm then simulates a gradient based on the random new position of the particle at the instance using the current delta potential well, which will push it either towards or away from the gbest value. This new position is then used as the new centre of each particle's respective delta potential well. Each iteration can hence be summed up as,

$$x = P \pm L/2ln(1/u) \tag{12}$$

where x is the new solution obtained, u is a uniform random number, and the movement is simulated by the obtained position in the delta potential well of each particle, where each particle's p will move with larger steps towards or away from the current solution based on the characteristic length of the well at that point given by l. However, there remains a probability that the new solution may not have been feasible or that it may have been less optimal than the prior position's solution due to the random nature of the obtained position at that instance. Hence, we shall add the rule,

$$\text{if } f(x_i) < f(p), \text{ then } p = x_i \tag{13}$$

which guarantees that there will be convergence and that the particles do not move away from their optimal and opposite to the gradient. Here, the update of the centre of the delta potential well for each particle is analogous to the update of its wave function as the two are directly related. Once the positions are updated, the algorithm then updates L and gbest as discussed earlier. After each iteration, each particle moves a little closer toward gbest while the particle at gbest also moves and possibly finds a better solution. This in turn leads to L and gbest being updated in case any of the particles come across better solutions. Eventually after several iterations, the particles and their corresponding pi values will converge toward a gbest solution.

Acknowledgment. We would like to thank the Department of Computer Science and Engineering at PES University, for encouraging and supporting us in writing this manuscript. We would also like to thank Dr. Snehanshu Saha for laying the groundwork through his endeavours in the field of Astro-Informatics and providing his expertise and knowledge during the course of the research. He emphasized the broad applicability of PSO techniques as shown in [11–15].

References

1. Sun, J., Feng, B., Xu, W.: A global search strategy of quantum-behaved particle swarm optimization. In: 2004 IEEE Conference on Cybernetics and Intelligent Systems, vol. 1, pp. 111–116, December 2004
2. Sun, J., Feng, B., Xu, W.: Particle swarm optimization with particles having quantum behavior. In: Proceedings of the 2004 Congress on Evolutionary Computation (IEEE Cat. No. 04TH8753), vol. 1, pp. 325–331, June 2004
3. Sun, J., Lai, C.-H., Wu, X.-J.: Particle Swarm Optimisation: Classical and Quantum Perspectives. Ed. by C.-H. Lai and F. Magoules. Chapman and Hall/CRC (2012)
4. Theophilus, A., Saha, S., Basak, S., Murthy, J.: A novel exoplanetary habitability score via particle swarm optimization of CES production functions. In: 2018 IEEE Symposium Series on Computational Intelligence (SSCI), pp. 2139–2147, November 2018
5. Bora, K., Saha, S., Agrawal, S., Safonova, M., Routh, S., Narasimhamurthy, A.: CD-HPF: new habitability score via data analytic modeling. Astron. Comput. **17**, 129–143 (2016). http://www.sciencedirect.com/science/article/pii/S2213133716300865
6. Hosseinpourfard, R., Javidi, M.M.: Chaotic PSO using the Lorenz system: an efficient approach for optimizing nonlinear problems (2015)
7. Ray, T., Liew, K.M.: A swarm with an effective information sharing mechanism for unconstrained and constrained single objective optimisation problems. In: Proceedings of the 2001 Congress on Evolutionary Computation (IEEE Cat. No. 01TH8546), vol. 1, pp. 75–80, May 2001

8. Benítez-Hidalgo, A., Nebro, A., García-Nieto, J., Oregi, I., Del Ser, J.: jMetalPy: a python framework for multi-objective optimization with metaheuristics (2019)
9. Méndez, A.: PHL's exoplanets catalog, November 2017. http://phl.upr.edu/projects/habitable-exoplanets-catalog/data/database
10. Eberhart, R., Kennedy, J.: A new optimizer using particle swarm theory. In: Proceedings of the Sixth International Symposium on Micro Machine and Human Science, MHS 1995, pp. 39–43, October 1995
11. Saha, S., Sarkar, J., Dwivedi, A., Dwivedi, N., Narasimhamurthy, A.M., Roy, R.: A novel revenue optimization model to address the operation and maintenance cost of a data center. J. Cloud Comput. 5(1), 1–23 (2015). https://doi.org/10.1186/s13677-015-0050-8
12. Sarasvathi, V., Iyenger, N.C.S.N., Saha, S.: QoS guaranteed intelligent routing using hybrid PSO-GA in wireless mesh networks. Cybern. Inf. Technol. 15, 69–83 (2015)
13. Saha, S., et al.: Theoretical validation of potential habitability via analytical and boosted tree methods: an optimistic study on recently discovered exoplanets. Astron. Comput. 23, 141–150 (2017)
14. Agrawal, S., Basak, S., Bora, K., Murthy, J.: A comparative analysis of the cobb-douglas habitability score (CDHS) with the earth similarity index (ESI), pp. 1775–1780, September 2018
15. Parsopoulos, K., Vrahatis, M.: Particle swarm optimization method for constrained optimization problem, vol. 76, pp. 214–220, January 2002

The System of Open Star Clusters Revisited

Priya Hasan$^{(\boxtimes)}$ and S. N. Hasan

Maulana Azad National Urdu University, Gachibowli, Hyderabad 500 032, India
priya.hasan@gmail.com, hasan.najam@gmail.com

Abstract. The system of open clusters is an excellent probe of the structure and evolution of the galactic disk. Their spatial, size, age and mass distributions provide valuable information on the cluster formation process. Present day astronomy is rich in data, and hence in this work, we attempt to build up a comprehensive statistical study of star clusters. This study is based on available catalogues, both homogeneous and inhomogeneous, to provide some useful insights on the evolutionary history of the system of open clusters and consequently, the galaxy. We find that the optimum size of a cluster for its survival is 3–4 pc. We also find that there exists a simple linear relationship between the age and the mean linear diameters of clusters and also with normalised reddening. Using the catalogues based on Gaia DR2 and other catalogs, we find, that the system of open clusters provides valuable clues to our understanding of the evolution of the galaxy. This system can be partitioned by k-means to get clusters in a statistical sense, which indicates possible cluster formation in the galaxy at different galactocentric distances and with different parameters. These suggests a combination of the scenarios of overall halo collapse and accretion to explain the formation of the disk of the galaxy. This method is proposed to be used for the study of external galaxies using catalogues of extragalactic clusters as it works well with the clusters of the Milky Way.

1 Introduction

Empirical research in astrophysics has seen a paradigm shift in recent years; it is now rich in data. Star clusters are the sites where star formation takes place and are ideal testbeds for the study of star formation, galaxy formation, evolution and dynamics. There have been numerous papers studying star clusters using photometry in optical and infrared bands [3,6,9,13,16]. In this work, we make a statistical study of star clusters based on both homogeneous and inhomogeneous catalogues and the package R [22] to provide some useful insights on the evolutionary history of open clusters and the Milky Way. It is hoped that the large number of open clusters for which parameters are now available can, in a statistical sense, lead to a deeper understanding of star clusters.

The plan of the paper is as follows: Sect. 2 describes the catalogues in our study and Sect. 3 describes the distribution of parameters in them. In Sect. 4,

© Springer Nature Singapore Pte Ltd. 2020
S. Saha et al. (Eds.): MMLA 2019, CCIS 1290, pp. 130–143, 2020.
https://doi.org/10.1007/978-981-33-6463-9_11

we discuss the relations between parameters. Section 5 describes the method of k-means clustering and the discussion and interpretation of our study is in the concluding Sect. 6.

2   Catalogues

[5] compiled a catalogue of open clusters in the Galaxy, the updated version (September 2017) has 2166 clusters. It updates the previous catalogues of [16] and of [19]. All the three catalogues are bibliographic in nature and contain compiled data from literature, thus presenting an inhomogeneous set of data. This catalogue is inhomogeneous since it has data in different photometric systems and varied methods of determination of parameters. Using a system of weights corresponding to the precision of the open cluster data, [10] studied the open cluster system by compiling a list of 694 entries of 421 clusters. [9] used the system of open clusters to study the development of the galactic disk. [25] studied the dependence of cluster diameters on various parameters based on the catalogue of [5].

Homogeneous samples of photometric data, coupled with uniform methods of data analysis are preferable to make statistical inferences on clusters. [15] published an updated catalogue of homogeneously estimated reddenings, distances from the Sun and ages for 425 open star clusters. [12] used the high-precision all-sky compiled catalogue ASCC-2.5 to derive a sample of 650 open clusters. On the basis of the combined spatial proper motion photometric membership they established uniform scales of spatial (angular dimensions), kinematic (proper motions and radial velocities) and evolutionary (ages) cluster parameters. The sample is complete within 850 pc, which provides unbiased parameters of the local cluster population and its evolution. The catalogue by [24] has 160 clusters for which parameters have been reobtained homogeneously by using CCD observations from literature. It has been used in this analysis as it has the vital parameter of mass, obtained homogeneously, which is not present in any of the other catalogues. [14] presented the Milky Way Star Clusters (MWSC) catalog of 3006 objects, the majority of them are open clusters, but also include associations and globular clusters. For each object they determined the exact position of the cluster centre, the apparent size, proper motion, distance, colour excess, and age.

In the case of star clusters, many uncertainties in determination of stellar parameters like reddening, distance, metallicity and stellar masses are minimised. The paper by [20] characterises the current status of knowledge on the accuracy of open-cluster parameters such as the age, reddening and distance. Hence, we expect, that in homogeneous samples there is similar accuracy in estimation of parameters, unlike that in compiled inhomogeneous samples.

Astronomical catalogues are said to be 'complete' if all the sources present have been listed in the catalogue. Completeness depends on the brightness of the sources and our capability of detection. The completeness of the catalogues we use are distance limited, implying that all sources within a certain distance have been detected. In Fig. 1 the solid line indicates possible completeness assuming that we have detected all clusters within 1 kpc.

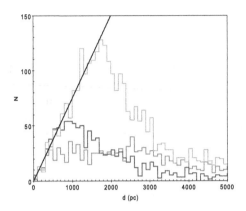

Fig. 1. Distribution in distance for catalogs by [5] in blue, [14] in green and [4] in red. (Color figure online)

The [5] catalogue is complete within 2–3 kpc [11]. The catalogue of [14] is complete within 1700 pc. The second Gaia data release DR2 [7,8] contains precise unprecedented astrometry at the sub-milliarcsecond level and homogeneous three-band photometry at the mmag level for about 1.3 billion sources, which can be used to characterize a large number of clusters over the entire sky. [4] provides a list of cluster stars based on membership and derived parameters, in particular mean distances and proper motions, for 1229 clusters, making use of Gaia data alone. This catalog is incomplete as the criteria used are very conservative, hence possible clusters at very close proximity have not been included.

For our analysis, we also added columns corresponding to galactocentric distance (GC) and linear diameters of the clusters in the catalogues. The GC has been calculated using the formula $R_{gc} = (R^2 + d^2 - 2Rd \cos l)^{1/2}$, where $R = 8.5$ kpc [1]. The reddening $E(B - V)$ has been replaced by normalised reddening [11] by the formula $E_n(B - V) = E(B - V)/r$ where r is the distance from the Sun in kiloparsecs.

3 Distribution of Galactic Longitude, Age, Reddening, Linear Diameters and Mass

The distribution of clusters in longitude (Fig. 2) is seen to exhibit maxima in the regions of active star formation at $l \approx 80^0$ (Vulpecula), $l \approx 125^0$ (Cassiopeia), $l \approx 205^0$ (Monoceros), $l \approx 240^0$ (Canis Major), $l \approx 285^0$(Carina) and $330^0 \leq l \leq 20^0$ (Sagitarrius). The deep minima occur in the obscured regions at $l \approx 50^0$, 155^0 and 195^0. The deepest minimum is in the region $l \approx 50^0$, where there is high obscuration due to dust. [25] infers from the reddening map of [11], evidence for

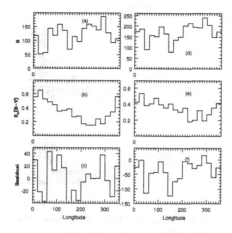

Fig. 2. Distribution of star clusters in Galactic longitude. Figures (a) and (b) are based on the [5] catalogue. Figures (d) and (e) show the [14] catalogue. The residual number of clusters ($N - N_{predicted}$) are shown in (c) and (f) to assess the effect of reddening on the number of clusters observed in each of the catalogues

a 4 kpc long dust arm that comes as close as 1.5 kpc at $l \approx 40^0$. We see that there is an excess of clusters between $200^0 \leq l \leq 300^0$. This could be an observational bias as the $E_n(B-V)$ is much smaller for that region. Hence, to test the effect of this bias, a plot of the mean $E_n(B-V)$ versus longitude (Fig. 2(b)) was made. From the plot of $E_n(B-V)$ versus number of clusters observed, we predicted the number of clusters to be observed for a given value of $E_n(B-V)$. The residuals $(N - N_{predicted})$ as a function of longitude are plotted in Fig. 2(c). A similar plots have been made with the homogeneous catalogue of [12] and is shown in Fig. 2(d), (e) and (f).

We observe a peak at $280^0 \leq l \leq 320^0$. This excess of clusters could be because the destructive tidal forces exerted by giant molecular clouds are weakest in the galactic anticenter direction [26]. Also the color excess $E_n(B-V)$ seems lower in that direction. It has also been suggested by [25] that this peak is because of clusters associated to the Canis Major dwarf galaxy (dCMa) ($l = 244^0, b = -8^0$). However, it was found by [21], that only Tombaugh 2 is physically located within the main body of dCMa and since this overdensity appears to be quite transparent to dust, only a few open clusters in that zone could have been missed.

In the [5] catalogue, there are 2102 clusters for which age has been determined and 299 of these lie within $280^0 \leq l \leq 320^0$. Figure 3(a) shows the age distribution of clusters in this region in the left panel and Fig. 3(b) shows the age distribution of clusters outside this region. We clearly see a peak in the region of $8 \leq log\, t \leq 8.5$. This shows a possible burst of cluster formation about 100 Myr ago which could be the cause of the high density of clusters in these longitudes. A similar peak is seen within $50^0 \leq l \leq 150^0$. This could also be tracing the spiral structure in the neighbourhood of the Sun.

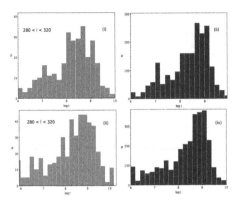

Fig. 3. Distribution in age for clusters within longitudes $280^0 - 320^0$ (red) and all clusters (blue) for the [5] and [14] catalogs. (Color figure online)

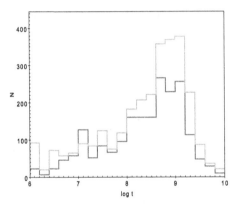

Fig. 4. Distribution in Age. The catalogues are represented by [5] in blue and [14] in green (Color figure online)

Figure 4 shows the distribution of age in clusters in the catalogues. The peak in age distribution at 100 Myr, appears in both catalogues, and agrees with [10]. [27] made a study of the age distribution and total lifetimes of clusters. He found the average lifetime of clusters to be 200 Myr.

Figure 5(a), (b) and (c) show the distribution of linear diameters in the catalogues of [5] and [14], where the peak diameters are 1.5 pc and the median diameters are ≈3 pc and 5 pc respectively. This implies that the optimum size of a cluster for its survival is close to the above value. This optimum size should also be related to the average mass of clusters, as a massive but small cluster will dissolve due to encounters between its stars and a larger one would break up under the galactic tidal field. Hence a fixed median size puts a constraint on the mass of a cluster.

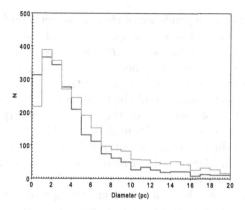

Fig. 5. Distribution in Linear Diameter. The catalogues are represented by [5] in blue and [14] in green (Color figure online)

4   Relations Between Parameters

4.1   Dependence on Longitude

To understand the relations between longitude on cluster parameters, we divided clusters into longitude bins of width of 40^0 in all the three catalogues and obtained the mean values of parameters in each bin with the exclusion of outliers.

Figure 6(a) shows the dependence of the normalised reddening $E_n(B-V)$ on longitude. We have plotted this to see the effect of observational bias in variable reddening for our next plots. As reported by [11], the [5] catalogue shows a sinusoidal dependence. A similar variation can also be observed in [12] and [24]. This pattern is related to the spiral structure in the vicinity of the Sun and hence is similar in these catalogues.

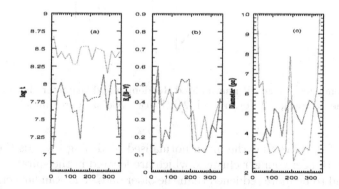

Fig. 6. Dependence of Age, Reddening and Cluster Diameter on Longitude. The data is from the catalogue [5] in blue and [14] in green (Color figure online)

Figure 6(b) shows the dependence of cluster diameter on longitude. It shows that the clusters in the anticenter region are found to be larger than those towards the centre. This could be due to two possible reasons: (i) Clusters towards the centre are smaller since they suffer greater galactic tidal forces and hence get disrupted earlier compared to those in the anticenter direction. (ii) The observational bias in determination of cluster diameter. Clusters in the anticenter region are found to be larger since the field star density in the anticenter direction is lower. The diameters obtained in the [14] catalogue are in general smaller than those in the [5] catalogue as the [14] catalogue is magnitude limited and hence does not consider the fainter members of clusters. However, we notice that both show a similar trend.

Figure 6(c) shows the dependence of cluster age on longitude. It can be noticed that the oldest clusters are in the anticenter direction in the [14] catalogue. It could be due to the larger survivability of clusters in the anticenter direction. The [5] catalogue which is a compilation, has younger clusters, while [14] is more effective in finding older clusters.

4.2   Dependence on Cluster Age

To understand the relations between evolutionary parameters we divided clusters into bins of $log\ t = 0.5$ in all the three catalogues and obtained the mean values of parameters in each bin with the exclusion of outliers.

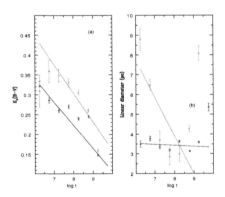

Fig. 7. Dependence of Age with normalised Reddening and Diameter. The data is from the catalogue [5] (red), [12] (cyan) and [24] (black). (Color figure online)

Figure 7(a) shows that the mean normalised reddening towards the clusters decreases with age: younger clusters, which are located in the spiral arms of the Galaxy, tend to be more reddened than the older ones. The standard error of the mean defined by $\frac{stddev}{\sqrt{N}}$ is shown in the plots. With a Pearson product-moment test we found a negative correlation of -0.89 and -0.93 with $p = 0.002$[1] for

---

[1] $p \leq 0.05$ implies a confidence level larger than 95 %.

both the [5] and [14] catalogues, respectively. Straight line fits have also made to the data. We found the linear relation

$$E_n(B - V) = m * (logt) + c$$

where $m = -0.058 \pm 0.01$, -0.07 ± 0.01 and $c = 0.69 \pm 0.09$, 0.88 ± 0.15 for the [5] and [14] catalogues respectively.

Figure 7(b) shows the relation of cluster diameter with age. As seen in the plot, the [5] catalogue shows a peak of large clusters at $log\ t = 7$. These, as we mentioned earlier are unbound clusters. In general, most clusters are 3–4 pc in size. This could be a selection bias, since these would be the easiest to observe clusters, In the case of [14], we clearly see that mean cluster diameters tend to decrease with an increase in age. This is because, as clusters age, they get mass segregated and start losing low mass stars at their outer regions. We found the linear relation

$$Diameter = m * (logt) + c$$

where $m = -1.93 \pm 0.78$ and $c = 19.35 \pm 5.96$.

5 K-means Clustering

After the above detailed exploration of the data, we carried out k-means clustering of the catalogues. The k-means algorithm is an algorithm to group n objects based on attributes into k partitions, $k < n$ [17]. It attempts to find the centres of natural $clusters^2$ in the data and looks for groups of objects which lie in close proximity to each other in the parameter space of these objects. It converges when it minimises the total intra-cluster variance, or, the squared error function

$$V = \sum_{i=1}^{k} \sum_{x_j \in S_i} (x_j - \mu_i)^2$$

where there are k clusters $S_i, i = 1, 2, ..., k$ and μ_i is the centroid or mean point of all the points $x_j \in S_i$. The main advantages of this algorithm are the simplicity and speed with which it can be run on large datasets.

In other words, the k-means cluster analysis criteria is employed to identify relatively homogeneous groups of cases based on selected characteristics using an algorithm which can handle large number of cases. For this algorithm we have to specify the number of clusters and also we have to specify the initial cluster center. With this information, we can classify the cases and then update the cluster centres iteratively. In our analysis, the initial cluster center and the number of iterations were taken to be 0 and 10 respectively. Then we specify a variable whose values are used to label case wise output (see Table 1 and 2).

---

^{2} $clusters$ in italics refers to clusters in the statistical sense and not physical star clusters.

K-means clustering was done for the [5] catalogue using the parameters: galactic longitude, galactic latitude, linear diameters, GC and age ($\log t$). Table 1 shows the centres of the *clusters* obtained for the [5] catalogue. Figure 8 shows how these *clusters* are spread in parameter space. These *clusters* are very clearly separated in GC. While the majority of *clusters* (in red, blue and black) are of age $\log t \approx 8$, there is an interesting *cluster* (in green) at negative latitudes -1.78^0 which corresponds to a $z = -1765$ pc with an age $\log t = 8.7$ and the largest GC ($GC = 15.46$ kpc) and comparatively large diameter of 9 pc. There is also a light blue *cluster* at a GC of 5.55 kpc and age $\log t = 8.28$.

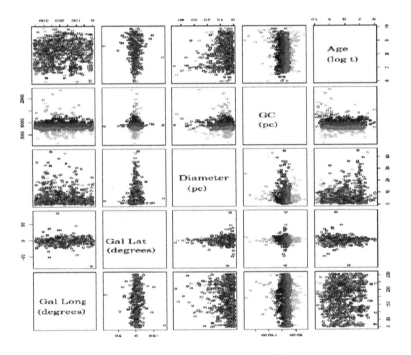

Fig. 8. K-means clustering [5]

Table 1. *Cluster* centres [5]

Gal longitude	Gal latitude	Diameter (pc)	GC (kpc)	$\log t$	No	Color
200	−0.50	6.6	5.55	8.28	73	Light blue
202	−1.50	5.6	7.70	8.08	299	Red
200	−1.20	5.8	8.97	8.09	374	Blue
182	−0.35	6.4	10.7	8.12	176	Black
223	−1.78	9.0	15.46	8.71	26	Green

K-means clustering was done for the incomplete [12] catalogue using the parameters: galactic longitude, galactic latitude, linear diameter of the core and cluster, GC and age ($\log t$). Figure 9 shows the results of the k-means clustering for the [12] catalogue. In this case, we have used the incomplete catalogue of [12], to highlight the differences between 'incomplete' and 'complete' catalogues. Table 2 shows the centres of the *clusters* obtained. The *clusters* differ by their GC, but do not differ very strongly in longitude, latitude and age. The *cluster* in red at high galactocentric distance has 26 young clusters with very large diameters and at a large GC of 10.75 kpc. In this table, we clearly see the effect of observational bias in catalogues. Clusters away from the Sun seem to have an average younger age, basically as they are more easily observable because of the presence of young stars. Hence, the clusters at GC of 3809 pc are young and the same applies to clusters in the anticenter direction.

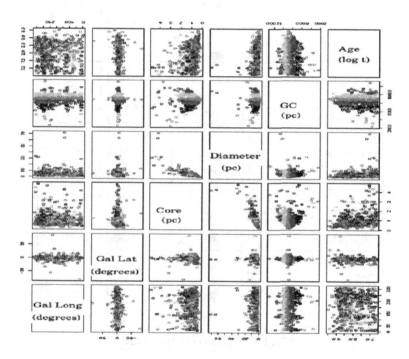

Fig. 9. K-means clustering [12]

K-means clustering was done for the [24] catalogue using the parameters: age ($\log t$), linear diameters, distance from the galactic plane (z), GC and Mass. Table 3 shows the existence of 4 *clusters* in the catalogue of [24]. Though this cluster is incomplete, the advantage with this catalogue is the homogeneously derived mass. In the case of clustering analysis with this catalogue, we have to be wary about the number of clusters in a *cluster*, as there is a large observational bias involved. However, the presence of *clustering* in the analysis is

Table 2. *Cluster* centres [12]

Gal. longitude	Gal. latitude	R_{core} (pc)	R_{clus} (pc)	GC (kpc)	$log\ t$	No	Color
211	−2.04	2.86	14.32	3.8	7.32	8	Lavender
231	−0.01	1.36	6.20	6.65	7.66	62	Black
223	−1.45	0.87	4.53	7.85	8.08	145	Blue
211	−1.76	0.85	4.84	8.68	8.17	181	Green
206	−1.04	1.07	5.20	9.5	7.96	97	Light blue
183	−0.78	1.92	11.79	10.75	7.41	26	Red

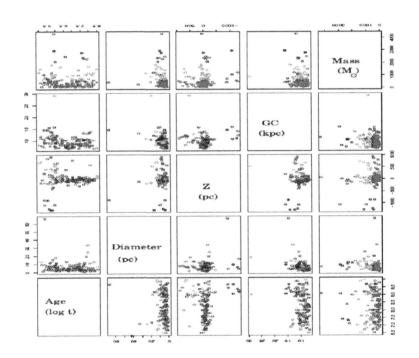

Fig. 10. K-means clustering [24]

Table 3. *Cluster* centres [24]

$log\ t$	Diam (pc)	z (pc)	GC (kpc)	Mass ($M_\odot$)	No	Color
8.12	4.78	43.26	9.33	267	91	Red
9.46	17.53	−1103	12.79	736	6	Blue
8.24	7.37	76.30	9.96	976	37	Green
8.08	7.91	−200	10.29	2707	8	Black

highly suggestive. As seen in Table 3, the *cluster* centres are described by age, diameters, z, GC and mass. The *clusters* are clearly separated in mass. There is a smaller *cluster* (blue) made of clusters of smaller masses. This group has clearly very negative $z = -1103$ pc, large $log\ t \approx 9.5$ and large diameters. There is also a *cluster* (in black) made of very massive clusters of mass 2707 $M_\odot$ at a distance of 200 pc below the galactic plane and diameter 7.9 pc. The origin of these *clusters* seems interesting. Since this is an incomplete catalogue, our observational bias would lead us to observe more younger clusters at larger distances. Hence, though this cluster has only 6 members, the true number must be much larger, since the catalogue is incomplete. Figure 10 shows how these *clusters* are separated in parameter space.

We should also note that unlike the previous two cases, clustering of the [24] catalogue was clearly in terms of mass and not GC, which indicates the significance of mass determinations in understanding clusters.

6   Discussion and Conclusions

Star clusters are very good indicators of the development of the galactic disk. In this paper, we have made a statistical study of the catalogues, both homogeneous and inhomogeneous, of star clusters. We studied the longitude, age, diameter and mass distribution of clusters in these catalogues. We find that the optimum size of a cluster for its survival is 3–4 pc. We have also studied the relation between parameters of clusters. The Pearson product-moment correlation test showed high negative correlations between age and normalised reddening $E_n(B - V)$ and age and linear diameters with a confidence level larger that 95%.

The review by [18] describes the two general categories of galaxy formation and evolution: overall halo collapse with subsequent disk formation and continued collapse under self-gravity and those in which accretion plays an important role. To study galaxy formation and to look for statistical *clusters* in our catalogues, we carried out k-means clustering of the data. This method has been described in detail in the earlier section. Its purpose is to look for groups within the sample which are closely related in parameter space. The results obtained have been described in detail. For the [5] catalogue, we found that while the majority of *clusters* are localised near the plane of the galaxy, at an average age of $\approx log\ t = 8$, there also exists a *cluster* of old star clusters of age log $t = 8.71$ and large diameters of 9 pc at a large GC. The bulk of clusters seem to be formed on the disk by the process of overall halo collapse. However, the group of old clusters could be formed by accretion as they are a at a distance of 1765 pc below the plane of the disk.

To highlight the differences between 'complete' and 'incomplete' catalogues, we have performed a k-means analysis of the incomplete [12] and [24] catalogues. We showed, how due to observational bias, the *clusters* located closer the Sun appeared to be older than those away in the analysis of the [12] catalogue. In the case of the [24] catalogue, we found that while the majority of clusters are at the plane of the galaxy, there were two smaller clusters away from the plane. There is a *cluster* made of very massive clusters with mass 2707 $M_\odot$ and age 100 Myr at a distance of 200 pc below the plane. There is also a group of still older clusters of age more than 1 Gyr (log t = 9.46) which is made up of very large clusters at a distance of 1103 pc from the galactic plane. The old clusters away from the galactic plane could survive as they are not perturbed by massive concentrations in the disk. These old clusters could not have been formed in the plane of the disk and then accelerated to their present positions as they would not survive the encounters necessary to move them [23, 28]. The presence of bound clusters away from the galactic plane are clear indicators of galaxy formation by accretion and infall onto the galactic disk. We find 26 such candidates from the [5] catalogue within 2–3 kpc of the Sun and 14 cluster candidates from the [24] catalogue which are clearly different compared the general cluster population and which are possibly formed due to accretion. A recent paper by [2] based on SDSS data supports a picture in which an important fraction of the stellar halo of the Milky Way has been accreted from satellite galaxies.

As a result of k-means clustering of these catalogues, we found that the catalogues could be grouped in *clusters* which indicate different episodes of star formation at different locations in the galaxy.

From our analysis, we conclude that the bulk of the data indicates galaxy formation by overall halo collapse, but, there is also evidence of formation by accretion from satellite galaxies from the existence of old clusters far from the galaxy plane. Studies based on more homogeneous data and other vital parameters (preferably mass) are required to a deeper understanding of the system of open clusters and the formation of the galaxy. Detailed studies of the interesting *clusters* found in the above study are also planned.

Acknowledgements. The authors also thank Hargopal Vajjha and C R Rao for comments and suggestions in statistics. Virtual observatory tools like VOSTAT and Topcoat have been used in the analysis.

References

1. Allen, C.W.: Astrophysical Quantities, 3rd edn. Athlone, London (1976)
2. Bell, E.F., et al.: The accretion origin of the Milky Way's stellar halo. **680**, 295–311 (2008). https://doi.org/10.1086/588032
3. Bonatto, C., Bica, E.: Detailed analysis of open clusters: a mass function break and evidence of a fundamental plane. **437**, 483–500 (2005). https://doi.org/10.1051/0004-6361:20042516
4. Cantat-Gaudin, T., et al.: A Gaia DR2 view of the open cluster population in the Milky Way. **618**, A93 (2018). https://doi.org/10.1051/0004-6361/201833476
5. Dias, W.S., Alessi, B.S., Moitinho, A., Lépine, J.R.D.: New catalogue of optically visible open clusters and candidates. **389**, 871–873 (2002). https://doi.org/10.1051/0004-6361:20020668
6. Friel, E.D.: The old open clusters of the Milky Way. **33**, 381–414 (1995). https://doi.org/10.1146/annurev.aa.33.090195.002121
7. Gaia Collaboration, et al.: The Gaia mission. **595**, A1 (2016). https://doi.org/10.1051/0004-6361/201629272

8. Hasan, P.: Gaia. Resonance **24**(4), 433–444 (2019). https://doi.org/10.1007/s12045-019-0796-2

9. Janes, K.A., Phelps, R.L.: The galactic system of old star clusters: the development of the galactic disk. **108**, 1773–1785 (1994). https://doi.org/10.1086/117192

10. Janes, K.A., Tilley, C., Lynga, G.: Properties of the open cluster system. **95**, 771–784 (1988). https://doi.org/10.1086/114676

11. Joshi, Y.C.: Interstellar extinction towards open clusters and Galactic structure. **362**, 1259–1266 (2005). https://doi.org/10.1111/j.1365-2966.2005.09391.x

12. Kharchenko, N.V., Piskunov, A.E.: The population of open clusters of the Galactic disc. Astron. Astrophys. Trans. **25**, 177–183 (2006). https://doi.org/10.1080/10556790600916863

13. Kharchenko, N.V., Piskunov, A.E., Röser, S., Schilbach, E., Scholz, R.D.: Astrophysical parameters of Galactic open clusters. **438**, 1163–1173 (2005). https://doi.org/10.1051/0004-6361:20042523

14. Kharchenko, N.V., Piskunov, A.E., Schilbach, E., Röser, S., Scholz, R.D.: Global survey of star clusters in the Milky Way. II. The catalogue of basic parameters. **558**, A53 (2013). https://doi.org/10.1051/0004-6361/201322302

15. Loktin, A.V., Gerasimenko, T.P., Malysheva, L.K.: The catalogue of open cluster parameters-second version. Astron. Astrophys. Trans. **20**, 607–633 (2001). https://doi.org/10.1080/10556790108221134

16. Lynga, G.: Open clusters in our Galaxy. Astron. Astrophys. **109**, 213–222 (1982)

17. MacQueen, J.: Some methods for classification and analysis of multivariate observations. In: Le Cam, L.M., Neyman, J. (eds.) Proceedings of the Fifth Berkeley Symposium on Mathematical Statistics and Probability, Berkeley, Califonia. University of California Press, vol. 1, pp. 281–297 (1967)

18. Majewski, S.R.: Galactic structure surveys and the evolution of the Milky Way. **31**, 575–638 (1993). https://doi.org/10.1146/annurev.aa.31.090193.003043

19. Mermilliod, J.C.: The database for galactic open clusters (BDA). In: Egret, D., Albrecht, M.A. (eds.) On-Line Data in Astronomy, pp. 127–+ (1995). ISBN 0-7923-3659-3. Astrophysics and Space Science Library, Vol. 203, p. 127 (Sep 1995)

20. Paunzen, E., Netopil, M.: On the current status of open-cluster parameters. **371**, 1641–1647 (2006). https://doi.org/10.1111/j.1365-2966.2006.10783.x

21. Piatti, A.E., Clariá, J.J.: The apparent overdensity of open clusters in the Canis Major overdensity. **390**, L54–L58 (2008). https://doi.org/10.1111/j.1745-3933.2008.00536.x

22. R Core Team: R: A Language and Environment for Statistical Computing. R Foundation for Statistical Computing, Vienna, Austria (2013). http://www.R-project.org/

23. Spitzer, L.J.: Distribution of galactic clusters. **127**, 17–+ (1958). https://doi.org/10.1086/146435

24. Tadross, A.L.: Morphological analysis of open clusters' propertiesI. Properties' estimations. New Astron. **6**, 293–306 (2001). https://doi.org/10.1016/S1384-1076(01)00061-6

25. van den Bergh, S.: Diameters of open star clusters. **131**, 1559–1564 (2006). https://doi.org/10.1086/499532

26. van den Bergh, S., McClure, R.D.: Galactic distribution of the oldest open clusters. **88**, 360–362 (1980)

27. Wielen, R.: The age distribution and total lifetimes of galactic clusters. **13**, 309–322 (1971)

28. Wielen, R.: The diffusion of stellar orbits derived from the observed age-dependence of the velocity dispersion. Astron. Astrophys. **60**, 263–275 (1977)

Genetic Bi-objective Optimization Approach to Habitability Score

Sriram Krishna[✉] and Niharika Pentapati

Department of Computer Science and Engineering, PES University,
Bengaluru, India
sriramsk1999@gmail.com, pniharika369@gmail.com

Abstract. The search for life outside the Solar System is an endeavour of astronomers all around the world. With hundreds of exoplanets being discovered due to advances in astronomy, there is a need to classify the habitability of these exoplanets. This is typically done using various metrics such as the Earth Similarity Index or the Planetary Habitability Index. In this paper, Genetic Algorithms are used to evaluate the best possible habitability scores using the Cobb-Douglas Habitability Score. Genetic Algorithm is a classic evolutionary algorithm used for solving optimization problems. The working of the algorithm is established through comparison with various benchmark functions and its functionality is extended to Multi-Objective optimization. The Cobb-Douglas Habitability Function is formulated as a bi-objective as well as a single objective optimization problem to find the optimal values to maximize the Cobb-Douglas Habitability Score for a set of promising exoplanets.

Keywords: Exoplanetary habitability score · Genetic Algorithm ·
Astroinformatics · Multi-objective optimization ·
Cobb-Douglas production function · Machine learning

1 Introduction

The search for life has been one of the oldest endeavours of mankind. But only recently have we acquired the capability to take even a step towards this lofty goal. With the first exoplanet discovered in 1991 [3], we have now reached a point where we have discovered over 4000 exoplanets. We have also taken steps in discovering if life exists on these planets through the use of various metrics such as the Earth Similarity Index [15] or the Cobb-Douglas Habitability Score (CDHS) [2]. These metrics take various planetary parameters as inputs and give us an intuitive understanding of the likelihood of life existing on these planets.

The Cobb-Douglas Habitability Production Function (CD-HPF) can quickly give us a score that is representative of the potential of habitability of an exoplanet. It takes in the Radius, Density, Escape Velocity and Mean Surface Temperature of a planet as inputs. All these inputs are in Earth Units (EU) i.e.

© Springer Nature Singapore Pte Ltd. 2020
S. Saha et al. (Eds.): MMLA 2019, CCIS 1290, pp. 144–157, 2020.
https://doi.org/10.1007/978-981-33-6463-9_12

the metric measurements of these parameters are divided by Earth's own measurements. Simply put, the values of any parameter of Earth in Earth Units is 1.

The Cobb-Douglas function was first developed in 1927 [4], seeking to mathematically estimate the relationship between workers, capital and goods produced. In its most standard form for production of a single good with two factors, it is written as

$$Y = AL^{\beta}K^{\alpha}$$

Where, Y is the total production, A is total factor productivity, L and K being the labour and capital inputs, and α, β being output elasticities of labour and capital respectively. The function itself is highly adaptable and has been utilized for various tasks like revenue models for data centers [14], frameworks for computing scholastic indicators of influence of journals [8] successfully.

The CDHS is calculated in a two-fold manner: by calculating the interior-CDHS using radius and density, and the surface-CDHS, by using escape velocity and surface temperature; the final score is computed by a convex combination of the two scores. Thus the function is formulated as a multi-objective optimization problem of the two scores.

Most optimization functions require the gradient of a function to minimize or maximize it. However, this can prove computationally costly and all functions are not differentiable, and even then, the derivative might not be smooth or continuous. In this paper, we use Genetic Algorithms, a class of gradient-free optimization functions, which are more widely applicable by virtue of them not requiring the derivative of the function to optimize it.

In the book, "On the Origin of Species" by Charles Darwin, he concluded that only those species survived who were successful in adapting to the changing environment and others died. He called this "Natural Selection" which has three main processes; Heredity,Variation and Selection. These involve species receiving properties from their parents, making variations to evolve and then being selected based on their adaptation to the environment for their survival. Along these lines, genetic algorithms [9] were introduced with five phases of process to solve an optimization problem. We create a initial population of randomly generated elements, known as solutions to the problem and then evaluate the correctness of the solutions using a fitness function which tells us how well the solution helps in optimizing the problem. Genetic Algorithms revolve around the twin principles of Exploration and Exploitation. There must be enough variety in the population to 'explore' the solution space which is usually vast, and on finding good solutions, the algorithm must 'exploit' these solutions and generate incrementally better solutions.

The typical Genetic Algorithm consists of 3 processes: Selection, Crossover and Mutation. In this paper, we use a modified version of a GA that combines the processes of Mutation and Crossover into one. This Proto-Genetic Algorithm is simpler to implement and understand while not compromising on performance.

We evaluate the 'fitness' of the population, that is to say we find the value of the function to be optimized using the members of the population, generate children using *one* parent and then test their fitness as well, choosing the best for the next generation. It is similar to the biological process of asexual reproduction where the child inherits all the traits from one parent alone. In this case, the child is generated from a Gaussian Distribution (as shown in Fig. 2) centered at the parent's value.

We illustrate the results of our algorithm on the set of Earth-like exoplanets that is the TRAPPIST system from the exoplanet catalog [12], hosted by the Planetary Habitability Laboratory at the University Of Puerto Rico at Acerbio.

2   Genetic Algorithms

A Genetic Algorithm (GA) is a meta heuristic which is based on the process of natural selection. It is a subset of the class of Evolutionary Algorithms which take cues from biological processes. They are most commonly used in optimization or search problems as they are capable of searching large combinatorial solution spaces to find globally optimal solutions.

Figure 1 indicates the pseudo-code of a typical Genetic Algorithm where a *population* of solutions are initialized randomly, given the constraints of a specific problem. The *fitness* of each solution is calculated, which is the value returned by the given function for that solution. Following which the genetic operators of Selection, Crossover and Mutation take place in order to create an incrementally better population. This process is repeated until a termination condition is met, such as a specified number of *generations*.

```
generate an initial random population
while iteration <= maxiteration
   iteration = iteration + 1
   calculate the fitness of each individual
   select the individuals according to their fitness
   perform crossover with probability pc
   perform mutation with probability pm
   population = selected individuals after
                       crossover and mutation
end while
```

Fig. 1. Pseudo-code of a typical Genetic Algorithm

2.1   Proto-Genetic Algorithm

In this paper, we have utilized a simpler version of the Genetic Algorithm. While GA's typically generate children using traits from both two parents, we have utilized a single-parent reproduction which is both crossover and mutation rolled into one. The best half of the population is selected and a single child is created for each parent. This child is created using a Gaussian Distribution centered at the parent, thus allowing for a mutation of sorts to occur. Due to the nature of the Gaussian Distribution, a small change is much more likely to happen than a drastic one, which reflects real life as well.

At the end of this process, we have a highly fit population. This algorithm is simpler to understand and implement but gives satisfactory results.

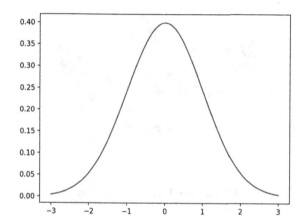

Fig. 2. The bell-shaped curve of the Gaussian Distribution

3 Implementation

3.1 Single Objective Optimization

Various test functions like Mishra, Rastrigin, Schaffer and others share many similarities [1]. All of them have 2 parameters and are highly multimodal. Thus, these functions serve as suitable benchmarks for GA.

We first initialize 2 sets of values for x and y, having populations of 200 values each. Our next step is to run the Genetic Algorithm. Here we choose to run the algorithm for 1000 generations, that is to say the processes of crossover, mutation and recombination take place 1000 times at the end of which we have solutions which are very close to the global minimum. The fitness measure here is of course the value of the function for the parameters x and y. After calculating the fitness for each pair we then choose the best pairs, i.e. the ones with the lowest fitness and then use them as the parents of the next generation. Generation of children is done using the Gaussian Distribution, allowing us to vary the children slightly in each generation. This is followed by checking the fitness of each child and arranging the children and the parents in order of their fitness. This weeds out all the parents who were not good enough and the children who were worse than their parents. Finally, we remake the population choosing the best of both the old and the new generation.

For example, the Rastrigin Function [11], a commonly used benchmark function used to test optimization algorithms due to its highly multimodal nature:

$$f(x,y) = 20 + x^2 + y^2 - 10(cos(2\pi x) + cos(2\pi y))$$

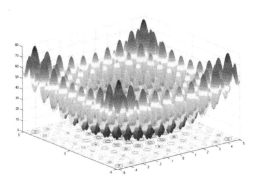

Fig. 3. The Rastrigin function

The Rastrigin function in Fig. 3 has a global minima of 0.0 with the domain being from -5.12 to 5.12. Thus, our algorithm generates 200 values of x and y in the given domain, which is the first generation of the algorithm. They are sorted according to their fitness and new children are generated from the best half of the population. Following this the population is remade by sorting according to fitness again and the second generation is created, with the members being slightly *fitter* than their parents. Table 1 compares the actual global minima and that obtained using GA for various test functions.

3.2   Constrained Optimization

These functions are also single objective optimization problems, however they are constrained. Whereas the previous batch of functions are only limited by the search domains, these functions have additional constraints. They tend to be more challenging to optimize.

Mishra's Bird Function displayed in Fig. 4 [10] has a global minima of -106.76 and the domain being from -10 to 0 for x and -6.5 to 0 for y. The function is given as:

$$f(x,y) = sin(y).e^{((1-cos(x))^2)} + cos(x).e^{((1-sin(y))^2)} + (x-y)^2$$

Table 1. Single objective optimization results

Test functions	Global minimum	
	Actual values	*GA values*
Easom	−1	−0.999
Rastrigin	0.0	0.0003
Ackley	0.0	0.009
Beale	0.0	0.0
Goldstein-Price	3.0	3.0001
Mishra No. 4	−0.199	−0.193
Cross-in-tray	−2.06	−2.06
Eggholder	−959.64	−959.27
Holder table	−19.208	−19.208
McCormick	−1.913	−1.913
Schaffer no. 4	0.292	0.292

In addition to minimizing this, the solutions must also not violate the additional constraint which is:

$$(x + 5)^2 + (y + 5)^2 < 25$$

We follow the same procedure as with single objective optimization, albeit making sure the solutions do not violate the constraints along with the upper and lower bounds of the domain. Children generated will be discarded if they do not satisfy the constraints. The standard deviation of the Gaussian Distribution goes on increasing to widen the search range if a large number of solutions are discarded. Table 2 lists different test functions along with their actual and GA obtained global minimum.

Table 2. Constrained optimization results

Test functions	Global minimum	
	Actual values	*GA values*
Rosenbrock (with a cubic and a line)	0.0	0.0009
Rosenbrock (disk)	0.0	0.0
Mishra's Bird	−106.76	−106.76
Townsend	−2.02	−2.02
Simionescu	−0.072	−0.0719

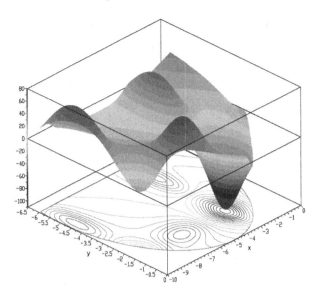

Fig. 4. Mishra's Bird function (Constrained)

3.3 Multi-objective Optimization

Whereas for single-objective optimization problems there exists a single solution which is the best value, no such solution exists for non-trivial multi-objective problems.

Multi-objective Optimization problems involve minimizing/maximizing more than one function simultaneously. If these functions are competitive i.e. minimizing one maximizes the other, it is not possible to find a single best solution. Instead we get a set of **non-dominating** solutions known as a **Pareto Front** (as shown in Fig. 6). In the absence of other information, each solution in the Pareto Front is equally valid and no solution can be said to be better than another.

Pareto fronts are based on the idea of dominance. If x, y are two solutions, then x is said to dominate y if

$$f_i(x)) \leq f_i(y)) \ \forall \ i = 1, 2, 3, ...k$$

In another words, the vector x is said to dominate y if and only if, $f(x) \leq f(y)$ for every single objective in the multi-objective optimization problem. We say that a vector of decision variables $x \in F$ is said to be Pareto optimal if no other vector $x \in F$ exists such that $f(y) \leq f(x)$. A multi-objective optimization consists of finding the best Pareto front for a given set of objectives.

There are various algorithms for multi-objective optimization. Indeed, one such algorithm, Particle Swarm Optimization (PSO) has already been used in solving the CD-HPF [17]. PSO have many advantages over GA [6] and hybrid PSO-GA have also been used in problems like intelligent routing [18] to great success.

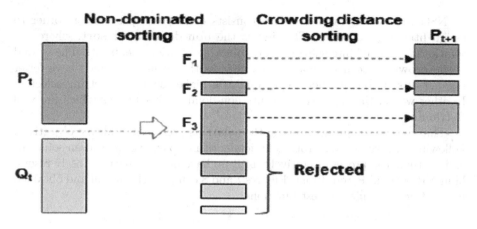

Fig. 5. Basics of NSGA-II procedure

When solving a multi-objective optimization problem using GA, a different approach must be taken. While in single objective problems we can directly compare function values as fitness and choose the best parents, the same cannot be done when we have multiple objectives to optimize. There are numerous algorithms such as MOGA [7], NSGA [16]. In this paper, we have used one of the most popular multi-objective optimization algorithms, NSGA-II [5].

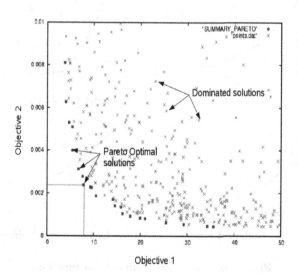

Fig. 6. A Pareto Front

NSGA-II as illustrated in Fig. 5 consists of two new processes in order to assign fitness to solutions. The first is the **non-dominated sort**, where the solutions are sorted into sets of non-dominating solutions i.e. fronts. The second is the **crowded-comparison**, which ensures that solutions which have fewer number of solutions in their vicinity have a higher chance of getting selected. In other words, this algorithm favours non-dominated solutions which are well distributed.

Thus we can assign a fitness to the solutions even with multiple objectives. Following this we use our proto-genetic algorithm to evolve the chosen solutions and continue the process iteratively until we have our population closely resembling the optimal Pareto Front. Figures 7 and 8 compare the actual and obtained pareto fronts of different test functions.

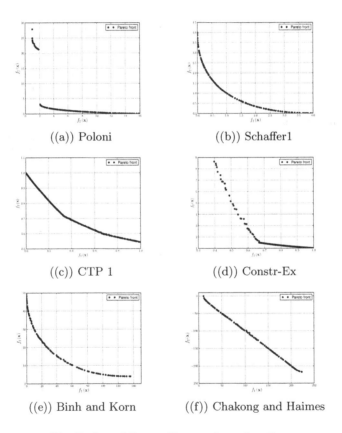

((a)) Poloni   ((b)) Schaffer1

((c)) CTP 1   ((d)) Constr-Ex

((e)) Binh and Korn   ((f)) Chakong and Haimes

Fig. 7. Actual Pareto Fronts of test functions

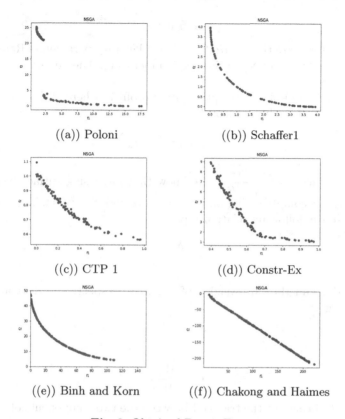

((a)) Poloni

((b)) Schaffer1

((c)) CTP 1

((d)) Constr-Ex

((e)) Binh and Korn

((f)) Chakong and Haimes

Fig. 8. Obtained Pareto Fronts

4 Cobb-Douglas Habitability Function

The Cobb-Douglas Habitability Function is given as follows:

$$Y = R^\alpha . D^\beta . V_e^\delta . T_s^\gamma$$

It can also be formulated as a bi-objective optimization problem for easy visualization and understanding. The Cobb-Douglas Habitability Score (Y) is divided into two components, CDHS-interior (Y_i) and CDHS-surface (Y_s). The CDHS is estimated by maximizing both Y_i and Y_s which are defined as follows:

$$Y_i = R^\alpha . D^\beta$$

$$Y_s = V_e^\delta . T_s^\gamma$$

These functions are subject to the constraints:

$$\alpha + \beta \leq 1$$

$$\delta + \gamma \leq 1$$

$$0 < \alpha, \beta, \delta, \gamma < 1$$

Where $\alpha, \beta, \gamma, \delta$ are the *elasticities* of the planetary parameters Radius, Density, Escape Velocity and Mean Surface Temperature. The quality of this model is well noted [13].

Thus we have a bi-objective optimization problem where we have to optimize $CDHS_i$ and $CDHS_s$ simultaneously.

$$max f(\boldsymbol{x}) = [Y_i, Y_s]$$

However, since $V_e = \sqrt{\dfrac{2GM}{R}}$, we know that increasing surface score is not possible without compromising on interior score and vice versa. Thus, as shown in [2], we use the following relationships:

$$V_e = \frac{\delta}{\alpha} \frac{W_R}{W_{V_e}} R$$

Where W_R and W_{V_e} are weights of R and V_e respectively. Rearranging the equation we get:

$$\delta = \alpha \frac{V_e}{R} C$$

where,

$$C = \frac{W_{V_e}}{W_R}$$

In order to bring out the trade-off between the two components of the Cobb-Douglas Habitability Score, we calculate δ from the other parameters, optimizing the variables α, β, γ and C.

We apply the aforementioned proto-genetic algorithm modified with NSGA-II on a set of planets from the exoplanet catalog hosted by the Planetary Habitability Laboratory at the University Of Puerto Rico, the TRAPPIST system. The results are shown in Table 3 with illustrations in Fig. 9.

5   Results

After testing on multiple exoplanets in the catalog, we found promising results similar to that of past approaches [13].

The Pareto fronts also show a trend where increase in one score is compensated for by decrease in the other. These complementary scores bring out the trade-off between Y_i and Y_s.

Table 3. Comparison of CDHS using GA with past approaches

Exoplanets	CDHS (2018)	CDHS (GA)
TRAPPIST-1 b	1.0410	1.3753
TRAPPIST-1 c	1.1589	1.2073
TRAPPIST-1 d	0.8870	1.0146
TRAPPIST-1 e	0.9093	0.9990
TRAPPIST-1 f	0.9826	1.0389
TRAPPIST-1 h	0.8025	0.9973
Proxima Cen b	1.08297	1.11909

The final score is calculated as the weighted linear combination of interior and surface score where the weights sum up to 1.

$$Y = w_i.Y_i + w_s.Y_s$$

$$w_i + w_s = 1$$

We set the weights w_i and w_s as 0.5 i.e. equal weights. Thus the calculated CDHS is the mean of the surface score and the interior score. With different weight pairs we get a range of habitability scores for each planet instead of a hard score, making the model more robust than other metrics.

In order to ensure the results are consistent, the CD-HPF was also solved as a single objective optimization problem i.e. its original form:

$$Y = R^\alpha.D^\beta.V_e^\delta.T_s^\gamma$$

Similar to other single-objective optimization problems, we generated populations of α, β, γ and δ and evolved them with the proto-genetic algorithm. The results were similar and establish the veracity of the multi-objective optimization approach as listed in Table 4.

Table 4. CDHS obtained using multi-objective and single objective optimization

Exoplanets	CDHS (multi-objective)	CDHS (single objective)
TRAPPIST-1 b	1.3753	1.3684
TRAPPIST-1 c	1.2073	1.2065
TRAPPIST-1 d	1.0146	1.0138
TRAPPIST-1 e	0.9990	0.9972
TRAPPIST-1 f	1.0389	1.0343
TRAPPIST-1 h	0.9973	0.9929
Proxima Cen b	1.11909	1.1158

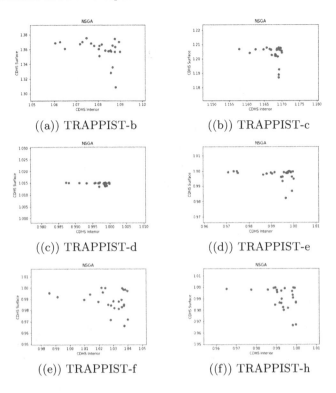

Fig. 9. The TRAPPIST system of exoplanets

6   Conclusion

In this paper, we have used a Proto-Genetic algorithm along with NSGA-II to calculate the best habitability scores for different exoplanets using the Cobb-Douglas Habitability Production Function. The optimizing capability of the proto-genetic algorithm was well established by testing on numerous benchmark functions of the single objective, constrained single-objective and multi-objective optimization types. Finally the algorithm was applied in calculating the habitability scores of promising exoplanets from the TRAPPIST system. The results were further verified using the single-objective optimization approach as well, establishing the merit of a genetic bi-objective optimization approach to habitability scores.

Acknowledgment. We thank our teachers who gave us this opportunity to work on something innovative and provided us a rich learning experience in the process. We thank Dr. Snehanshu Saha for assistance with this endeavour and guiding us in this project. And finally, we would also like to show our gratitude to the PES University for allowing us to go beyond our coursework and work on a project with practical applications.

References

1. Back, T.: Evolutionary Algorithms in Theory and Practice: Evolution Strategies, Evolutionary Programming, Genetic Algorithms. Oxford University Press, Oxford (1996)
2. Bora, K., Saha, S., Agrawal, S., Safonova, M., Routh, S., Narasimhamurthy, A.: CD-HPF: new habitability score via data analytic modeling. Astron. Comput. **17**, 129–143 (2016)
3. Cochran, W.D., Hatzes, A.P., Hancock, T.J.: Constraints on the companion object to HD 114762. Astrophys. J. **380**, L35–L38 (1991)
4. Coma, C.W., Douglas, P.H.: A theory of production. In: Proceedings of the Fortieth Annual Meeting of the American Economic Association, vol. 139, p. 165 (1928)
5. Deb, K., Pratap, A., Agarwal, S., Meyarivan, T.: A fast and elitist multiobjective genetic algorithm: NSGA-II. IEEE Trans. Evol. Comput. **6**(2), 182–197 (2002)
6. Eberhart, R.C., Shi, Y.: Comparison between genetic algorithms and particle swarm optimization. In: Porto, V.W., Saravanan, N., Waagen, D., Eiben, A.E. (eds.) EP 1998. LNCS, vol. 1447, pp. 611–616. Springer, Heidelberg (1998). https://doi.org/10.1007/BFb0040812
7. Fonseca, C.M., Fleming, P.J., et al.: Genetic algorithms for multiobjective optimization: formulation discussion and generalization. In: ICGA, vol. 93, pp. 416–423. Citeseer (1993)
8. Ginde, G., et al.: ScientoBASE: a framework and model for computing scholastic indicators of non-local influence of journals via native data acquisition algorithms. Scientometrics **108**(3), 1479–1529 (2016)
9. Holland, J.H., et al.: Adaptation in Natural and Artificial Systems: An Introductory Analysis with Applications to Biology, Control, and Artificial Intelligence. MIT Press, Cambridge (1992)
10. Mishra, S.K.: Some new test functions for global optimization and performance of repulsive particle swarm method. Available at SSRN 926132 (2006)
11. Mühlenbein, H., Schomisch, M., Born, J.: The parallel genetic algorithm as function optimizer. Parallel Comput. **17**(6–7), 619–632 (1991)
12. Planetary Habitability Laboratory: Exoplanet catalog (2019). http://phl.upr.edu/projects/habitable-exoplanets-catalog/data/database
13. Saha, S., et al.: Theoretical validation of potential habitability via analytical and boosted tree methods: an optimistic study on recently discovered exoplanets. Astron. Comput. **23**, 141–150 (2018)
14. Saha, S., Sarkar, J., Dwivedi, A., Dwivedi, N., Narasimhamurthy, A.M., Roy, R.: A novel revenue optimization model to address the operation and maintenance cost of a data center. J. Cloud Comput. **5**(1), 1–23 (2015). https://doi.org/10.1186/s13677-015-0050-8
15. Schulze-Makuch, D., et al.: A two-tiered approach to assessing the habitability of exoplanets. Astrobiology **11**(10), 1041–1052 (2011)
16. Srinivas, N., Deb, K.: Muiltiobjective optimization using nondominated sorting in genetic algorithms. Evol. Comput. **2**(3), 221–248 (1994)
17. Theophilus, A., Saha, S., Basak, S., Murthy, J.: A novel exoplanetary habitability score via particle swarm optimization of CES production functions. In: 2018 IEEE Symposium Series on Computational Intelligence (SSCI), pp. 2139–2147. IEEE (2018)
18. Sarasvathi, V., Iyengar, N.C.S.N., Saha, S.: QoS guaranteed intelligent routing using hybrid PSO-GA in wireless mesh networks. Cybernet. Inf. Technol. **15**(1), 69–83 (2015)

Machine Learning Based Analysis
of Gravitational Waves

Surbhi Agrawal[1], Rahul Aedula[2(✉)], and D. S. Rahul Surya[3]

[1] Department of Computer Engineering, ADYPU, Pune, India
[2] Department of Computer Science, University of Colorado,
Boulder, USA
rahulaedula95@gmail.com
[3] Sprinkler Technologies, Timnath, USA

Abstract. Gravitational waves has been a serious subject of study in the modern day astrophysics. Where on one end the strain produced by gravitational waves on matter could be practically studied by Laser Interferometers such as LIGO, the strain generated by celestial bodies on the other end a priori obtained by numerical relativity in the form of waveforms. It is often the case that these waveforms are only used to study the properties of black holes. This article tries to extrapolate such methodologies to weaker celestial bodies for the primary purpose of adding a new dimensionality in the prudent realm of possibilities. There is a necessity to approach such studies from a statistical perspective. Utilizing the combination of Statistical and Machine Learning tools not only assist in analyzing data effectively but also aid in creating a generalized computational model.

Keywords: LIGO · PyCBC · Gravitational waves · Numerical relativity · Regression · Classification

1 Introduction

Gravitational waves (GW) are formally defined as the ripples in the space-time curvature [1]. They are the direct consequence of a body with mass accelerating through space. Scientifically, it occurs when the mass quadruple moment changes with time. Einstein's introduction of the theory of general relativity first planted the seeds of inception for Gravitational Waves, now after a span of 100 years with help of the extraordinary efforts of LIGO and Virgo their existence has been confirmed. On February 11th 2016, LIGO announced its first discovery of Gravitational waves corresponding to two black holes of 36 and 29 solar masses merging. So far three such events of black hole mergers have been recorded.

Gravitational waves are detected as strains in laser interferometers when it passes through it. These strains given by, $h(t) = \Delta L/L$ where ΔL is the change in length, is often converted into numerical relativity based waveforms to better understand the source. Numerical relativity (NR) is a subsidiary of general

© Springer Nature Singapore Pte Ltd. 2020
S. Saha et al. (Eds.): MMLA 2019, CCIS 1290, pp. 158–175, 2020.
https://doi.org/10.1007/978-981-33-6463-9_13

relativity. It is often employed to study cosmological entities such as black holes and neutron stars. The principles of NR regarding GW however remain the same regardless of the type of entity which is being studied. This leads to a postulation that NR can be extrapolated to other entities particularly weaker entities such as Exoplanets etc. The primary problem with employing NR directly to weaker entities is that the complex nature of NR algorithms and its reliance on a converging point gives it an enormous overhead in terms of time complexity and therefore is ill-suited to obtain results quickly. There is a significant necessity to find an optimized solution for generating NR waveforms, as NR waveforms give valuable insight about the source. There have been attempts made to computationally increase the speed of NR equations such as PyCBC which still have considerable overheads which will be addressed in depth in the upcoming sections.

Aside from the computational aspects of generating NR based waveforms for GW, there is also a need to use GW as a feature which can supplement the existing cohort of information regarding Exoplanets. This direction deals with the extrapolation of properties of Gravitational Waves to weaker celestial bodies like Exoplanets. Using the GW information to corroborate the classification of Exoplanets into their respective mass classes. So far no such models exist, during the progression of this article a precursory model is introduced which covers the aforementioned basis.

2   Basic Properties and Features of Gravitational Waves

2.1   Physical Properties

In the context of physical nature, Gravitational waves cause the stretching and squeezing of matter in space, it also distorts time around the object causing a slowing and speeding up of time. GW also has polarization similar to light they are (i) Plus and (ii) Cross type polarizations respectively. This polarization is caused due to the precession of the binary in-spiral pair. Gravitational Waves being ripples in Space-time propagate at the speed of light. The necessary conditions required for the propagation of GW is

$$\lambda << R$$

where λ is the wavelength of the GW and R is the Radius of Curvature (ROC) of the background space-time. Other properties such as absorption and dispersion are negligible in Gravitational Waves.

2.2   Wave Characteristics

Gravitational Waves can only be studied and discerned by their waveforms and not just the strain alone. It is not possible to map GW as a figure or a picture. Waveforms contains the details of the source. They can hold many attributes such as mass of the binary pair, GW frequency etc. Normally to decide these attributes it has to be compared with NR simulations which as mentioned before takes a long time even with the capabilities of existing super computers. To circumvent this problem, numerical relativity approximates are used. Numerical relativity approximates are simulated waveforms which are done in a weak field paradigm. Under the assumption that the entities in question are not moving fast enough or to better phrase it moving slower than the speed of light. This application of Einstein's equations in a weak field paradigm is also called Post Newtonian Expansion.

$$h(f) = \frac{1}{r} M_{ch}^{5/6} f^{-7/6} exp(i\psi(f)) \tag{1}$$

$$\psi(f) = 2\pi f t_c - \phi_c - \frac{\pi}{4} + \frac{3}{128}(\pi M_{ch} f)^{-}5/3 \tag{2}$$

Here t_c is time at coalescence, ϕ_c phase at coalescence and M_{ch} is chirp mass which will be discussed in the upcoming sections. The h is a first order approximation of the strain. Both these equations correspond to the frequency domain. A few adjustments can be made regarding the location of the source but for a basic scenario these formula can be generalized. These formula give insight into how these source attributes can be obtained by the wave.

2.3   Existing Computational Approximation Methods

As previously mentioned NR takes a lengthy amount of time to approach generalized solutions. To ease the arduous computational task approximation is used. This provides an optimized solution with the help of the aforementioned Post Newtonian Expansion methods. A common type of waveform is that of phenomenological waveform or phenom waveform. These type of waveforms have shown a very successful rate in mimicking NR waveforms as closely as possible. It has some inaccuracies as it has come to be expected, because of the approximated nature of its generation but over all efficiency has been significantly high compared to most approximates. One such existing methods which uses phenom based waveforms is PyCBC. **PyCBC** - an open source python implemented stable module could be used to obtain the theoretical gravitational waveforms for specific input parameters such as the masses, lower frequency and so on. This powerful module gives an optimized solution to theoretical equations by means of Bayesian Belief Networks and computes different types of gravitational waveform such as SEOBNR, TAYLOR and many more. The one problem with PyCBC is that even though it is an optimized solution it can only give results for black holes and other heavy entities like neutron stars. GW cannot be studied effectively regarding other lesser mass entities like Exoplanets etc. which are also in in-spiral with their corresponding star. Through the course of this paper we will be extrapolating the approach used for black holes on multiple Star-Exoplanet Systems. The aforementioned SEOBNR(Spin Effective One Body Numerical Relativity) is a phenom based waveform which will be the primary focus as we progress with this article (Fig. 1).

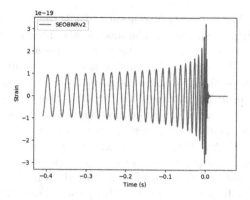

Fig. 1. PyCBC generated SEOBNR waveform

2.4 Proposed Computational Approach

The problem lies in creating a flexible and elegant solution by discerning both aspects, the waveform and the germane physics required to correlate the idea to other weaker entities. On one hand there has to be a lucid interpretation of the waveform, that is to say there should be clear understanding of how the trend of the waveform changes with respect to different parameters which influence it and on the other hand the proposed astrophysical model should not only validate but also make inferences by a supervised learning procedure. The way to go about achieving these goals is to approach the problem in two simultaneous subroutines.

These proposed mechanisms are:

– Regression Analysis
– Classification

Regression Analysis deals with discerning the trend of the SEOBNR waveform. It tries to correlate the various parameters that are inclusive of the generation of these waveforms and tries to have a pellucid grasp on how it can be made computationally efficient and deals with the process of extrapolation outside the domain of its limiting factors. Classification aims to propose a new model to group these entities in question with the help of GW. Not only does such a model not exist so far but also it helps in validating and correcting the regression results in a peculiar way. The upcoming sections of this article delves deep into these subroutines and aids in surfacing a new computational model which is created by mending these approaches.

3 Regression Analysis

The previous section speaks about the PyCBC module and why it is marked to be of utmost importance in the study of GW. It is a robust module to perform tasks at various levels studying GW, one such task is the generation of GW based on numerical relativistic equations. The module also presents diverse waveforms to pick for study, ranging from the Taylor series representation to the SEOBNR and many more. An important observation here is - the numerical relativistic equations or the theoretical equations involved behind the generation of these waveforms are intrinsic by nature. This intrinsic property results in a lot of computational overhead while using the module for some specific values supplied as part of functional requirements such as the masses of the celestial objects for the generation of GW. An other observation is that, beyond certain limit of the input parameters such as the masses of celestial bodies in-spiral, PyCBC fails to compute values and generate waveforms, though in reality a similar in-spiral system would naturally generate a GW. This paper aims to introduce few basic concepts of ML, Regression and analyze their contribution to bring down the computational overhead and extend the domain of the input parameters while trading off the accuracy of the waveform generated by a small amount. This trade off should be meager, shouldn't interfere producing a result too deviated from the theoretical result. The most interesting phase of in-spiral is that of the Coalescence. As discussed in the previous sections about the in-spiral, there is a certain period of time after which they gain rapid acceleration and spin vehemently about each other. This is followed by both of the celestial bodies colliding and thus merging into a single body which is a common phenomenon in binary Black holes. The start of this merger is marked by the coalescence. The peak amplitude of the GW happens at the coalescence and plays a vital role to study the properties of celestial bodies involved in generating this peak amplitude.

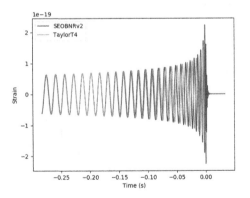

Fig. 2. A sample PyCBC Waveform

The above Fig. 2 depicts a simple PyCBC generated waveform for the parameters $m_1 = 10$, $m_2 = 10$, $spin1z = 0.0$, $delta\_t = 1.0/4096$ and $f\_lower = 60$. The peak amplitude during coalescence (h_0) as seen from the waveform occurs at $Time = 0.00$.

To understand how the peak amplitude varies with various masses of the celestial bodies in-spiral, a huge dataset was created with a mixture of masses, recording the peak amplitudes during coalescence from the PyCBC module. In fact, the peak amplitude was recorded against the chirp mass which is given by,

$$M_{ch} = \frac{(m_1 m_2)^{3/5}}{(m_1 + m_2)^{1/5}} \tag{3}$$

where m_1, m_2 correspond to the masses of the celestial bodies with m_1 being the mass of the more massive body among the two. The generation of the dataset involves a lot of time as the generation of the waveform even for a single input requires a lot of time. Thus, parallelization was used with the help of Multi-processing module in python to build this dataset using all cores of the CPU. The dataset created comprises peak amplitudes during coalescence (SEOBNR) of celestial bodies recorded for different values of masses m_1, m_2 and lower frequency f as follows:

Parameter	Range
m_1	$10 - 99$
m_2	$10 - m_1$
f	$\{35, 40, 45, ...60\}$
spz	0
del	$6.103515625E - 05$

It has to be noted that the peak amplitude during coalescence for any input lower frequency got to be equal in magnitude, but for PyCBC generated waveforms the peak amplitudes are not strictly equal though approximately equal. Thus, the lower frequency is also considered while generating the data set. For the preliminary analysis on the relation between the peak amplitude during coalescence against the masses of the celestial bodies, a scatter plot is plotted with these parameters.

From the above Fig. 3 it could be observed that the scatter plot approximates to a linear curve. The x-axis corresponds to M_c while the y-axis corresponds to $h_0 \times 10^{19}$. The linear model of this data set could be imagined to be

$$h = \beta_0 + \beta_1 M_{ch} + \epsilon \tag{4}$$

In the above equation, β_0 and β_1 is some coefficient and ϵ the error. The blue line in the above equation is a linear fit which could be calculated using the Sum of squares method, minimizing the sum of the squared errors. Thus, for the linear fit $\hat{h} = \hat{\beta}_0 + \hat{\beta}_1 M_{ch}$, we obtained the parameter $\hat{\beta}1 = 0.25464428$. The term $\hat{\beta}_0$ or the intercept is zero, since it is logical that when $M_c = 0$, $h_0 = 0$. It is now possible that we make use of this model to obtain h_0 for other celestial bodies such as the exo-planets, which could in turn be used for the classification of exo-planets as discussed in the later sections of this paper.

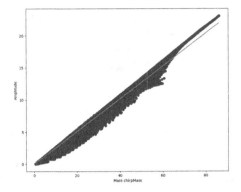

Fig. 3. Regression on amplitude peaks

4   Complete Waveform Generation

In the previous sections we have obtained a model to predict h_0 for any given M_{ch}. In this section we discuss about an approach which could be used in the generation of complete waveform. The approach taken here does not generate the exact waveform but an approximation of the waveform. Generating the complete waveform would result in a lot of GW characteristics which have potential applications in many fields of astronomy. As we can observe from the PyCBC waveform, output for SEOBNR, it is evident that the amplitude versus time waveform is a chirp equation whereby the frequency increases with time.

To generate a waveform with unique characteristics, the waveform could be identified with the amplitude, the frequency and its phase. Now, to generate the GW waveform envelope we need to have at least two characteristics, the amplitude and the frequency. The reason being that, the peak amplitude during coalescence always occurs at the time $= 0$. It could be observed that the amplitude peaks versus time of a gravitational wave could be split into two parts. By observation, it could be concluded that in the first part of the GW the time increases exponentially with the amplitudes peaks. i.e., the relation between amplitude peaks and time is with time t and amplitude h follows:

$$t = e_a.e^{h.e_b} + e_c \tag{5}$$

The coefficients e_a, e_b and e_c of this model cannot be obtained analytically and hence we resort to use approximation algorithms to obtain these values. For obtaining these values we make use of the *curve_fit()* method inside *scipy.optimize*. The initial guesses for these constants were set to $(-1, -1, -1)$ Now, h_{max} or the envelope could be obtained from t just using

$$h_{max} = \left(\frac{1}{e_b}\right).ln(\frac{t - e_c}{e_a}) \tag{6}$$

The fit of the curve looks as depicted in Fig. 4.

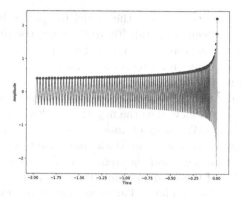

Fig. 4. PyCBC generated SEOBNR waveform

For the fit in the curve, the points marked in the red color, forming an envelope above are the predicted peaks. The curve fits pretty well for the amplitude peaks, but as we could see, reversing the equation presents two issues viz. the first thing the domain error and the second thing that the predicted values increase rapidly after a certain interval of time. It could be defined that the time interval upto which neither of the issues occur as the pre-coalescence phase or the non-coalescence phase. The time interval after which either of the two issues occur could be defined as the coalescence phase, where by the current exponential model fails to fit the model and hence we would have to go with another model which could be a model such as:

$$t = x_a.h^{x_b} \tag{7}$$

Reversing the above equation for amplitude gives,

$$h_{max} = \left(\frac{t}{x_a}\right)^{\frac{1}{x_b}} \tag{8}$$

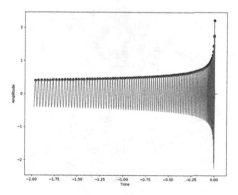

Fig. 5. PyCBC generated SEOBNR waveform

It could be seen from the fit that the model fits perfectly for the values of time after the non-coalescence period. Figure 5 shows the fit of the amplitude peaks for the time after the non-coalesce period or during the coalescence period.

The points in black color, forming an envelope towards coalescence are the predicted peaks during the coalescence phase.

An envelope around the graph peaks could be obtained by using the models for both the positive peaks as well as the negative peaks. Now in order to obtain this envelope for all the GW for given masses and lower frequency, we need to run a regression analysis on how these Model parameters viz. e_a, e_b, e_c, x_a and x_b vary with respect to masses and the frequency. The images - Fig. 6, 7, 8 and 9 show 3D scatter plot of the Model parameters with respect to chirp mass and frequency for the non-coalesce phase. The points green in color are the predicted points while the points in blue color are the observed points. From Fig. 6 and Fig. 7 it can be inferred that the distribution of these parameters against M_{ch} and f is exponential.

It could be visually confirmed that the relation of e_a with chirp mass and the frequency is exponential, in fact it has a score of roughly about 98.5%. Similarly, e_b and e_c follow an exponential distribution. The image of x_a against M_{ch} and f shows a 3D scatter plot of the Model parameters with respect to the chirp mass and the frequency for the coalescence phase.

It could be visually confirmed that the relation of x_b with chirp mass and the frequency also follows an exponential fit. But the relation of x_b with respect to the chirp mass and frequency is slightly scattered. Thus, for this we could take a density estimate. The reason for this being that the scatter plot becomes uniform as the plot approaches toward the lower masses and a lower frequency but still the regression line passes through the mean density of the scatter.

After we are able to obtain an estimate of these model parameters, we could again obtain an envelope of the amplitude for the time of a GW for a given chirp mass and a given frequency by just using a version of gradient descent on the equations up to the lower frequency.

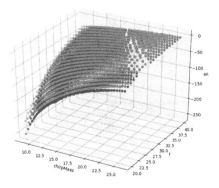

Fig. 6. e_a against M_{ch} and f

The procedure for construction of the envelope goes as follows:

1. Compute h_0
2. Initial time,

$$t = x_a.h^{x_b}$$

$$\frac{dt}{dh} = x_a.x_b.h^{x_b-1}$$

3. Use $\frac{dt}{dh}$ for new time,

$$t_{new} = t_{old} - \frac{dt_{old}}{dh}$$

4. Compute h_{new}

$$h_{new} = \left(\frac{t_{new}}{x_a}\right)^{1/x_b}$$

5. Compute f_c (current frequency)

$$f_c = \left|\frac{1}{t_{new} - t_{old}}\right|$$

if $f_c <= f_{lower}$: stop else if:

$$h' = \frac{1}{e_b} \ln\left(\frac{t - e_c}{e_a}\right) > h_{new}$$

then goto step 3 else: continue

6.

$$\frac{dt}{dh} = e_a.e_b.e^{e_b.h}$$

7.

$$t_{new} = t_{old} - \frac{dt_{old}}{dh_{old}}$$

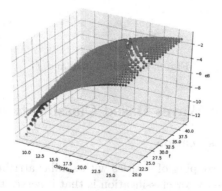

Fig. 7. e_b against M_{ch} and f

8.

$$h_{new} = \frac{1}{e_b} \ln \left(\frac{t - e_c}{e_a} \right)$$

9. if

$$f_c = \left| \frac{1}{t_{new} - t_{old}} \right| <= f_{lower}$$

then stop else goto step 3.

Now that the envelope of the wave would be generated the next question would be on identifying the points at which the envelope of the wave has to be considered to roughly approximate the original wave. For this a simple scatter plot between the chronological order of positive peaks of amplitude versus time gives us a hyperbolic curve as shown in Fig. 10.

Thus, going with a hyperbolic fit, the right lower part of the equation

$$\frac{t^2}{a^2} - \frac{i^2}{b^2} = 1 \tag{9}$$

where t corresponds to the positive time and i corresponds to the chronological order or index of the positive peaks. The next section deals with how the predicted amplitude during coalescence could be applied for exo-planets and help in their classification.

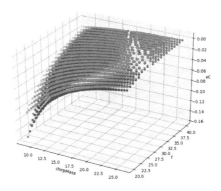

Fig. 8. e_c against M_{ch} and f

5   Gravitational Waves Based Classification

5.1   Need for Classification

Classification is the grouping of entities with similar attributes under a defined class label. The necessity for classification is that it serves two purposes, it validates the astrophysical model and corroborates all the assumptions made so far in extrapolating the model outside its usual domain, but also it helps in a subtle form of correctable operations which corrects any types of inconsistency that may occur

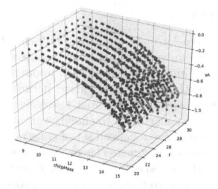

Fig. 9. x_a against M_{ch} and f

due to the regression analysis. Primarily it is used to bolster the claims made so far that the model can indeed be flexible enough to operate outside the domain. The center of focus in this section is dealing with the implications of extrapolating the already existing GW concepts to weaker entities, followed by some computational evidence that supports the usage of the GW concepts in this manner. The weaker entities under scrutiny in this scenario are Star-planet system of Exoplanets. The reason to choose this particular set of celestial objects is because they follow the inspiral mechanisms which is similar to black holes. The revolution of an Exoplanet around their respective star which is considerably far away behave as a weak pair of binary coalescing black holes. The term "weak pair" here describes the reduced magnitude of mass compared to that of a black hole. The assumption is that coalescence takes place a few million years later rather than quickly like that of two merging black holes. Newtonian mechanics dictate that the gravitational orbits are stable and once a celestial body such as a planet enters into orbit it remains in revolution forever. But the introduction of Einstein's general relativity showed otherwise, indeed there is a decay in the gravitational orbits over time and this decay of energy is emitted out in the form of Gravitational Waves. This Gravitational waves can be observed in planets the problem being that the wave itself is too weak to be detected by any conventional detectors like LIGO. The reason being that a planet system would emit a GW due to the motion of masses which are many times weaker than that of black holes, particularly LIGO is currently more tuned to extract GW information for only black holes. This means that to practically detect these waves from such small sources would probably take more calibration on LIGO's end but that doesn't say anything about interpreting the theoretical results. These theoretical results can assist in creating a more suitable model to correlate weak mass entities with GW. The waveforms are well discerned in the previous section Regression Analysis, the results generated for each unique wave can be used as aids to help create a more pellucid model. Results such as the maximum peak amplitude of the supposed coalescence, the GW frequency and the known mass of the planets all help in creating a better classification model (Table 1).

5.2 Classification Overview

Gravitational waves have a lot of factors which influence them but none more prominent than mass. As mentioned in the previous sections the necessity of Chirp Mass in extracting important information about the source is essential. When GW is so prominently influenced by mass, it can also be used as an aid to help better sort the masses of the entities themselves. The mass class is a parameter defined by astronomers used to classify various Exoplanets by virtue of their mass. These masses have a defined set of constraining values which segregate them to their corresponding mass classes. The numeric mass class are the integer allocated values of the respective mass class done for easy graphic representation. The key idea here is to relate the different masses of these Exoplanets to the Gravitational waves they may produce while orbiting their corresponding star. A supervised machine learning model can be proposed on the grounds of training data with already confirmed masses along with their respective mass classes to that of the supposed Gravitational Wave information. The Regression Analysis section shows the various wave information that can be extracted by using various predictive modeling techniques. Although to reach this coalescence point where there is an immense burst of GW from the corresponding potential merger it would take a considerable amount of time, at least a few billion years. However, this potential maximum peak amplitude (GWPeakAmp) can be used as a feature in classification of these Exoplanets. This is because the maximum peak still denotes a unique point in the NR waveform. It can be used to uniquely identify a binary pair based on the Chirp Mass, thereby making it a trainable feature in the classification model. Another feature which is used to train the classification algorithm is the Gravitational Wave frequency (GWFrequency). The frequency of the wave itself can tell us lot about the source, it can correlate to the orbital dynamics of the binary pair of the star-planet system of the Exoplanet. The features which determine the outcome of Machine learning based classification are as follows:

– GWPeakAmp
– SunMassSU
– PlanetMassSU
– GWFrequency

These features provide a rudimentary extension to the already existing mass class grouping mechanism, so by associating them we can not only increase the accuracy of the classification but also provide a new model, a catalog which encompasses Gravitational Waves along with the existing features in the Exoplanet Catalog.

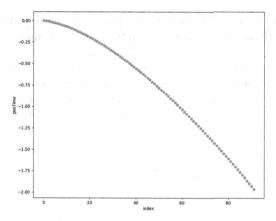

Fig. 10. Index vs time (positive peaks)

Table 1. Planet mass classes

Mass class	Numerical mass class (NMC)
Jovian	1
Terran	2
Superterran	3
Subrterran	4
Neptunian	5
Mercurian	6

5.3 Exoplanet Catalog Dataset

The application of Gravitational Waves makes it a requisite to use the most reliable catalog to extrapolate the idea. One such catalog is the Planetary Habitability Laboratory Exoplanet Catalog (PHL-EC) by the University of Puerto Rico. Although this dataset is known for its study for habitability of planets the usage of the data set in this scenario is very different. The PHL-EC data set is being used only to derive the orbital mechanics related information such as orbital frequency and the corresponding masses of the planets. The mass classes are also a part of this robust dataset. Although, there might exist some inconsistencies when masses are taken into account, such as compression of gravity and the scaling of mass with volume. All these factors might cause a skew in the data set for some of the mass classes. The dimensionally reduced distribution figure Jovian class for example contains the most varying range of possible planets, but there may be cases in the data set where some of the observations are misclassified. For this reason there is no hard limit set on the mass classes. This problem can be overcome by our classification strategy by incorporating the Gravitational Waves as a feature giving more clarity and distinction to the mass classes. Another recognizable problem with the dataset is that it is unbalanced. This may cause problems with the algorithm as it causes a bias in the ML model. The reason a bias like this might exist is because if there are more samples present in the training data which belong to a particular class compared to the other classes, this will cause the algorithm to identify the majority sample class with more efficiency compared to that of the minority sample class. This type of imbalance problems can be resolved in various over-sampling or under-sampling techniques. The next section will cover these methods in more detail (Fig. 11).

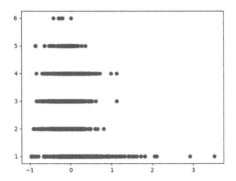

Fig. 11. Dimensionally reduced data distribution

5.4   Classification Strategy and Algorithm

In the previous sections, the discussion was primarily focused on the idea and the requisites that are needed for the implementation of the idea. This section deals with the algorithmic details and methodologies. One of the primary aims in this approach is to be able to classify accurately into any mass class. This means a clear distinction of mass class has to be established. As mentioned before there are some small amount of inconsistencies in the dataset. The process of selecting the classification algorithm must take that into consideration this nature of data and should provide an improvement. The appropriate algorithm which satisfactorily tends to these nuances is **Random Forests**. Random Forests being based on decision trees, finds the boundaries of the classes. This is needed as previously stated, all the mass classes have to be uniquely identified and should provide the best accuracy for each class. Under ideal circumstances such an algorithm would be more than perfect to reliably complete the task. But the aforementioned problem of imbalance tends to create biases within the data. The two methods which are present to tackle such problems are (*i*) Over-sampling and (*ii*) under-sampling techniques. Over-sampling techniques include procedures to artificially generate data to fill the gaps so as to minimize then imbalance, where as Under-sampling includes using only part of the data to which all the classes are balanced, without using the entire dataset. For this specific GW based classification problem, the better choice was over-sampling.

Another reason for performing oversampling is that future data which has yet to be observed and recorded has not been accounted. The generation of synthetic data will help test the model in more robust conditions. There are many well known oversampling techniques such as ADASYN etc., but for this problem, Synthetic Minority Over-Sampling Technique (SMOTE) was the superior method. Other methods such as kernel density estimation, ADASYN were also evaluated, but SMOTE provided the most agreeable values. The requisite for the synthetic data is to maintain it's integrity with that of the original PHL-EC data. SMOTE provides values in an aggregated range and also manages to maintain the magnitudes of the parameters within domain constraints. Thus solving the imbalance problem. The next course of action was to apply Random Forests, to the already balanced data. The reasons for choosing Random Forests over Decision Trees is because Random Forests uses a significant amount of voting based conclusions as compared to that of Decision Trees. It runs a bagging based routine by using a large number of de-correlated Decision Trees to classify a predicted class. This course of operations is highly suitable for the GW data and it's associated mass classification as it meticulously examines the feature space to make better judgments over which mass class to choose.

5.5   Classification Performance Metrics

The accuracy metrics decide how efficiently the model has performed over a set of constraining factors. In the case of classification, it could be simply defined as the accuracy involved in predicting the correct class for a given set of parameters. The results show various metrics such as True Positive Rate (TPR) or Sensitivity and True Negative Rate (TNR) Specificity etc. These metrics give valuable insight into how each class is being treated and gives a more lucid interpretation of all the nuances of the data. The following results displayed consists metric scores of both classification without using SMOTE and classification after using SMOTE (Table 2).

Table 2. Overall Metrics: PM = Performance Metrics; Original Data = OD; Original + Synthetic Data = OSD

PM	OD	OSD
Overall accuracy	89.4505%	84.9932%
95% CI	(0.8625, 0.9211)	(0.8325, 0.8651)
Kappa	0.8488	0.8184

The scores mentioned in Table 3 show the overall performance of the Random Forest classifier. As the results indicate there is a drop in the overall accuracy in the results of the SMOTE generated data. This is because the classifier is trained without any imbalance and hence showing a decrease in accuracy. But this classifier trained without imbalance is more robust compared to that of the classifier which trained only on the original data. The higher accuracy score of the original data can only be attributed to the skew in the data. Because of imbalance the classifier tends to recognize class 1 (Jovian) more than any other class. Also, in the actual data set the number of planets in the class 6 or Mercurian category are very less which is visible in its peculiar class wise scores. This also factors into the classification algorithm's accuracy.

Table 3. Class wise performance metrics

NMC	Specificity		Sensitivity		Accuracy		F1 score	
	OD	OSD	OD	OSD	OD	OSD	OD	OSD
1	0.99	0.98	1.00	0.99	0.99	0.99	0.99	0.98
2	0.89	0.88	0.99	0.97	0.97	0.95	0.92	0.83
3	0.70	0.73	0.96	0.92	0.92	0.89	0.74	0.71
4	0.78	0.54	0.92	0.93	0.90	0.87	0.69	0.56
5	0.88	0.91	0.99	0.99	0.99	0.97	0.91	0.93
6	0	0.90	1.00	0.99	0.99	0.99	0.000	0.98

The class-wise scores give a better insight on how the classifier is handling each class separately. Understanding the variations in Specificity and Sensitivity are key in discerning how to efficiently boost up the classification model. As shown in Table IV the class wise scores, the over all performance of the synthetic model improved compared to it's counterpart. This goes to show that even though overall accuracy is higher, the robustness of model might not be prominent. The model built from using SMOTE and then applying Random forests might score less in the overall accuracy (not by much), the robustness of the model is far more superior. It has validated that it can handle the data it may have to process in the future and show promising results.

6   Conclusions

Gravitational waves and its significance has just started emerging to the forefront. This article has taken steps in a creative direction of applying these physical phenomenon to other weaker entities. The proposed computational model serves as a rudimentary approach to a far more perplexing design. The strongest aspect of this model is the use of Machine Learning and Statistical tools to not only ease the understanding of this complex phenomenon but also making it efficient to operate with it. The perspective of Data Science and Machine Learning is that of elegance. Physicists already understand a vast amount about Gravitational Waves. Data Science not only enhances that plethora of knowledge but also gives a unique outlook. An eloquent solution to generalize one of the most arcane paradigms of the universe. The progress made in this article is a stepping stone for more elaborate models yet to be made. The understanding of Gravitational waves is absolutely vital in advancement of sciences. It's correlation with some of the most enigmatic entities like black holes make it all the more reason to delve deep to discern them. In search of these answers the proposed approaches suggested in this paper is the understanding of these waveforms and extrapolating the information learned from these trends to far outside the domain. Our understanding of the waveform and how it behaves over variation of various parameters such as mass and frequency has been enhanced. The application of this knowledge outside the usual domain is a step on the creative side of the paradigm. It led us to developing an efficient way of producing waveforms, and also a new method for classifying Exoplanets based on the GW released by the star-planet binary system. Ultimately developing a computational model comprising the two which enhance each other. The application of this model lies in optimization algorithms to generate waveforms and also catalog extraction which incorporates GW with Exoplanets.

7 Future Scope

This proposed computational model is the first of many approaches that will unfold. With better understanding, a more refined model can be created by considering a set of more elaborate factors. The proposed classification model may use more discernible features and make the classification more robust. Better calibration of the LIGO sensors can help in physically validating the proposed waveforms. GW reveals more details about the universe everyday, a new off domain application can be brought to existence by correlating the right idea with accurate physics.

References

1. Abbott, B.P., et al.: Observation of gravitational waves from a binary black hole merger. J. Astrophys. Phys. Rev. Lett. **116**(6), 061102 (2016)
2. Abbott, B.P., et al.: Properties of the binary black hole merger GW150914. J. Astrophys. Phys. Rev. Lett. **116**(24), 241102 (2016)
3. Devine, C., Etienne, Z.B., McWilliams, S.T.: Optimizing spinning time-domain gravitational waveforms for advanced LIGO data analysis. Class. Quantum Gravity **33**(12), 125025 (2016)
4. Berti, E., et al.: Inspiral, merger, and ringdown of unequal mass black hole binaries: a multipolar analysis. Phys. Rev. D **76**(6), 064034 (2007)
5. Martynov, D.V., et al.: Sensitivity of the advanced LIGO detectors at the beginning of gravitational wave astronomy. Phys. Rev. D **93**(11), 112004 (2016)
6. Berti, E.: The first sounds of merging black holes. arXiv preprint arXiv:1602.04476 (2016)
7. Khan, S., et al.: Frequency-domain gravitational waves from nonprecessing black-hole binaries. II. A phenomenological model for the advanced detector era. Phys. Rev. D **93**(4), 044007 (2016)

Thermal Suitability Scheme: Habitability Classification of Exoplanets

Suryoday Basak[(⊠)]

The University of Texas at Arlington, Arlington, USA
suryodaybasak@gmail.com

Abstract. In this paper, a metric for estimating the potential habitability of exoplanets, called thermal suitability score (TSS) is developed based on machine learning (ML). As compared to prior literature, the TSS ascertains habitability by using a sign – positive for potentially habitable, and negative for non-habitable – and a number indicating the extent to which an exoplanet is habitable or non-habitable. The TSS is used on the data provided in the University of Puerto Rico's Planetary Habitability Laboratory's Exoplanets Catalog (PHL-EC).

Keywords: Machine learning · Exoplanets · Habitability · Thermal suitability score

1 Introduction

A majority of the interest in exploring exoplanets is the possibility of the existence of life on planets other than earth. For millennia, scientists and philosophers have pondered over this possibility. Today, the rate at which exoplanets are being discovered is rapidly increasing, fuelled by technological and methodological advances. In this era of rapid discovery, it is imperative to develop methods that can summarise the properties of planets such that interesting planetary samples may be found and studied easily. Habitability metrics such as the biological complexity index (BCI) and the planetary habitability index (PHI) were developed with this intent. Further developments based on econometric modeling done by [1–3] have expanded the range of techniques that may be used to quantify properties of exoplanets.

In this paper, a habitability metric that is driven by machine learning is presented. Machine learning (ML) is a set of tools and ideas that involve statistical estimation and inference in a way that ensures that a machine may be able to make an intelligent decision. The typical families of machine learning are divided into supervised and unsupervised methods. In supervised machine learning, the training data consists of feature-target pairs, $(\mathbf{x}, y)$, where $\mathbf{x}$ is an input to the system and y is the outcome for x that is learned by the algorithm. In a way, in supervised learning, an optimal mapping between $\mathbf{x}$ and y is learned, for all $\mathbf{x}$ in the input space. In unsupervised learning, the training data does not have

© Springer Nature Singapore Pte Ltd. 2020
S. Saha et al. (Eds.): MMLA 2019, CCIS 1290, pp. 176–184, 2020.
https://doi.org/10.1007/978-981-33-6463-9_14

a target to which the input is mapped; rather, an algorithm tries to discover patterns that are inherent to the data and try to find underlying categories or discriminations in the absence of a target label. Classification and regression are supervised learning tasks, whereas clustering is an unsupervised learning task.

The metric that is proposed is called the thermal suitability score (TSS). After ascertaining the effectiveness of ML algorithms for the automatic classification of exoplanets [2] a method to quantify the potential of a planet to be habitable is developed, based on surface temperature alone, by developing an entirely data-driven metric. This is called the Thermal Suitability Score (TSS) because it is developed by using the mean surface temperature of a planet (and features extracted from the surface temperature). The motivation to do this lies in the fact that exoplanets that would be of greatest interest to the astronomical community would be those that are closest in terms of their properties to earth; the most important characteristic of Earth is the existence of water on the surface, and the surface temperature of a planet is thus an optimistic indicator of the possibility of surface water: if a planet is too hot (surface temperature over 100 °C), then it is unlikely that liquid water would exist on its surface as it could have possibly evaporated; likewise, if a planet is too cold, (surface temperature below 0 °C), then it is likely that even if water is present on the surface, it is frozen and would not be able to support life as it is on earth.

2   Method

2.1   Understanding Surface Temperature Based Discrimination of Exoplanets Based on Machine Learning

The Thermal Suitability Score (TSS) is a score which, in addition to providing a notion of similarity to Earth in terms of surface temperature, provides a habitability classification of an exoplanet. The formulation of this method is based on support vector machines (SVMs). As a part of this method, two classes are used based on the optimistic sample of potentially habitable exoplanets by PHL [4]. The two classes are those of *potentially habitable* and *non-habitable* exoplanets. The TSS is determined by first finding the maximum separating *hyperplane* between the classes in the data, which acts as a discriminator and using the distance from the hyperplane as the key characteristic. The metric is then developed by normalizing this distance by dividing by the distance of the Earth's feature vector from the hyperplane.

The goal of this model is to find a score that can instantly help us discriminate between potentially habitable and non-habitable planets by finding one boundary between two classes in the data. This is a hybrid approach where a model outputs a number and a sign, the number indicating similarity to earth, and the sign indicating the class. In this light, surface temperature (S. Temp) is one of the only features which can be used to develop the metric because S. Temp (and the related features of flux and distance from parent star) are the only features based on which the *habitable* and *non-habitable* samples are *reasonably* linearly separable.

2.2   Formulation of the Optimization Problem

The support vector machine (SVM) quadratic optimization problem [5], which is the basis of the TSS, is given as:

$$\min_{\lambda} \frac{1}{2}\lambda^T(yy^T K)\lambda - \lambda$$
$$\text{subject to } -\lambda \leq 0,$$
$$y \cdot \lambda = 0$$
(1)

where y is the list of class labels corresponding to samples in the data, λ is the set of Lagrange multipliers, and K is the Gram matrix, which is given as:

$$K(x_1, ..., x_m) = \begin{bmatrix} x_1 \cdot x_1 & x_1 \cdot x_2 & \cdots & x_1 \cdot x_m \\ x_2 \cdot x_1 & x_2 \cdot x_2 & \cdots & x_2 \cdot x_m \\ \vdots & \vdots & \ddots & \vdots \\ x_3 \cdot x_1 & x_3 \cdot x_2 & \cdots & x_3 \cdot x_m \end{bmatrix}$$
(2)

where x_i represents the ith sample in the data.

After the optimization problem has been solved and the support vectors have been found, the weight and bias: the variables w and b are determined by:

$$w = \sum_{i=1}^{m} \lambda_i y_i x_i$$
$$b = \frac{1}{m} \sum_{i=1}^{m} y_i - w \cdot x_i$$
(3)

where m is the number of samples in the dataset.

The features used in this method are of the following form:

$$x = (T, |T - 1|)$$
(4)

where $|\cdot|$ represents the absolute value function and T is the surface temperature in Earth units. Together, these two features give us a data representation of the surface temperature and the similarity of the surface temperature of the planet to that of the Earth's. This implies that the Earth's feature vector is $(1, 0)$ and the consequence of this is that in the feature space, the distance of Earth from the maximum separating hyperplane is the maximum. In addition to this, as a consequence of the discrimination done by the hyperplane, the output of the method is positive for potentially habitable samples (and non-habitable samples whose surface temperatures are near the hyperplane) and it is negative for non-habitable samples. Let the distance of the Earth from the maximum separating hyperplane be represented as d. The final expression for the score is thus given as:

$$TSS = \frac{y \cdot (w \cdot x + b)}{d}$$
(5)

2.3   Feature Extraction

The absolute value of the difference between the surface temperature of the exoplanet and the surface temperature of the Earth is used as a parameter in addition to the value of the surface temperature of the exoplanets, whose value in EU is 1. Thus, the model inputs become ordered pairs of the type $(T, abs(T-1))$. A data implication of extracting a feature this way is that the value of $abs(T-1)$ is 0 for Earth, an aspect central to the scoring mechanism of the model.

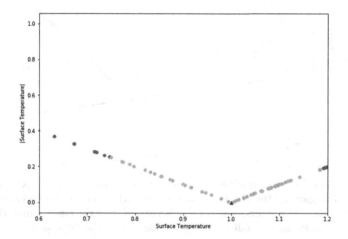

Fig. 1. Plot of S. Temp vs absolute value of (S. Temp - 1). In EU, 1 is the S. Temp of Earth, and in this graph, it is represented by the green triangle at $(1, 0)$. The points in orange represent the optimistic sample of potentially habitable exoplanets and the points in blue represent non-habitable planets. Since the non-habitable set is more expansive, only the points in the vicinity of the habitable samples are plotted. We see that there is minor overlap near the boundaries of the classes. (Color figure online)

From a physical viewpoint, we now have a representation of a planet's surface temperature in comparison to Earth. From a computational viewpoint, we have an added dimension in the dataset which will help us effectively separate and score the potentially habitable planets from the non-habitable planets using a single hyperplane in a 2D space.

2.4   Overlap Between Classes

In the SVM formulation for a linearly inseparable dataset [5], there is a minimization of a classification error which provides the *best* boundary between the classes in the data. However, in this method, we do not want to find a best-case boundary of separation, but would like to be inclusive of the habitable samples which are manually labeled by the PHL-EC as we consider them to be reliable points of judgment of habitability. Hence, we find the convex hull of the habitable samples and exclude the non-habitable samples within this convex hull prior to finding a separating hyperplane. By doing this, we get a perfect separation between the two classes, and we proceed to find the optimal separating hyperplane.

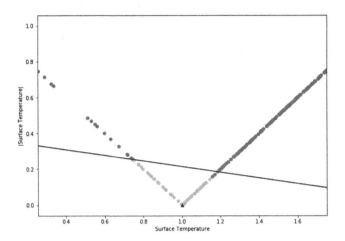

Fig. 2. A graphical depiction of the optimal separating hyperplane (black line) determined after disregarding the non-habitable points within the convex hull of samples of the habitable class. The consequence of this is noticed near the class boundaries, where a few non-habitable samples fall on the habitable-side of the hyperplane.

2.5 A Geometric Interpretation of the Method and Proof of Maximum Value

Let the feature vectors be denoted as $(T, g(T))$, where

$$g(T) = \begin{cases} L_1 = -T + 1, & \forall T < 1 \\ L_2 = T - 1, & \forall T > 1 \end{cases} \tag{6}$$

$g(T)$ is nothing but an expansion of the absolute value function. L_1 and L_2 may be considered as two lines which intersect at $(0, 1)$. Let the separating hyperplane be denoted by H. As we know that a separating hyperplane in a 2D space is a line, L_1, L_2 and H may be considered to form a triangular region if the angle made b H with respect to L_1 and with respect to L_2 is zero. This is proven below.

Let the angle between H and L_1 and L_2 be θ_1 and θ_2 respectively. If $\sin\theta_1 > 0$ and $\sin\theta_2 > 0$, then we can say that H is not parallel or collinear with respect to L_1 and L_2 respectively. If $\sin\theta_1$ and $\sin\theta_2$ are both greater than 0 then H intersects with both L_1 and L_2.

The slope of L_1 is -1 and the slope of L_2 is 1. Let the angle between L_1 and L_2 be given by θ_3. Then,

$$\theta_3 = \arctan \frac{-1 - (1)}{1 + (-1)(1)}$$

$$= \arctan \frac{-2}{0} \tag{7}$$

$$= \frac{\pi}{2}$$

Thus, considering the points of intersection of H with L_1 and L_2 being $A = (x_1, y_1)$ and $B = (x_2, y_2)$, and considering $O = (1, 0)$, we can assert that a triangular region is formed by OAB. Also, from Eq. 7, we know that $\triangle OAB$ is a right angled triangle.

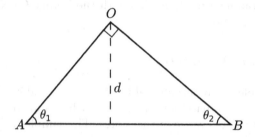

Fig. 3. Geometric representation of the problem in a 2D space. AB is a segment from the optimal separating hyperplane, $A = (x_1, y_1)$, $B = (x_2, y_2)$ and $O = (1, 0)$. d is the distance of $(1, 0)$ (which is Earth's feature vector) from the separating hyperplane.

Let us consider the side OA. Here, O and A are the end points. On this line segment, we know that point O is the greatest distance away from point A. This is proved by contradiction.

Let us assume that on this line segment, O is not at the greatest distance away from A. Let there be a point O' on OA such that it lies between O and A and is at a greater distance away from A than O. We know that for any point K between O and A,

$$|OA| = |AK| + |KO| \tag{8}$$

Hence, for the point O',

$$|AO| = |AO'| + |O'O|$$
$$\Rightarrow |AO'| = |AO| - |O'O| \tag{9}$$
$$\Rightarrow |AO'| < |AO|$$

However, this contradicts the premise that there can be a point O' on the line segment OA for which $|O'A| > |OA|$. This implies that on line segment OA the greatest distance between any two points on the line is the distance between the endpoints O and A and this is the length of the line. This further implies that there can be no point on OA apart from O for which $|A'O|\sin\theta_1 > |AO|\sin\theta_1$.

Keeping in mind the geometric representation as shown in Fig. 3, we see that the distance between O and AB, which actually represents the segment of the hyperplane included in the triangular region $\triangle OAB$, is given by $|AO|\sin\theta_1$. The same can be proven by taking into consideration side OB instead of OA.

Thus, in the context of the problem, in the feature space, Earth is the furthest away from the maximum separating hyperplane and the distance of any planet with feature vector not equal to $(0,1)$ from H will be less than that of Earth.

Thus, finally, the conditions that arise which allow this model to be used as a metric is that the separating hyperplane should not be collinear or parallel to any of the sides of the triangular region formed by $\triangle OAB$. While solving the problem, we find that this condition is satisfied by the data. We find w $= [-392.011, -2487.989]$, $b = 923.011$ and $d = 531.0002$. The solution of the problem was programmed in Python3.6 with the library CVXOPT, which is a library for convex optimization.

3   Results and Discussion

The TSS of a sample of potentially habitable and non-habitable exoplanets are presented in order to compare how the metric behaves for different exoplanets, and to understand the relevance of the scores. The samples of TSS is presented in Table 1.

Table 1. The TSS (2.1) of various samples are presented.

Potentially habitable exoplanets			Non-habitable exoplanets		
P. Name	S. Temp	TSS	P. Name	S. Temp	TSS
TRAPPIST-1 d	1.01527	0.91713	TRAPPIST-1 b	1.37674	-1.04331
TRAPPIST-1 e	0.90416	0.62172	TRAPPIST-1 c	1.20799	-0.12806
TRAPPIST-1 f	0.79757	0.20096	TRAPPIST-1 h	0.63125	-0.45554
TRAPPIST-1 g	0.75035	0.01456	Kepler-519b	1.97257	-4.27495
ProximaCenb	0.91632	0.66969	MOA-2010-BLG-328Lb	0.33403	-1.62874
Kepler-186f	0.77430	0.10913	OGLE-2005-390Lb	0.32187	-1.67671
Kepler-705 b	1.00555	0.96987	Wolf1061d	0.56042	-0.73513
K2-72e	1.10555	0.42749	YZCetb	1.64861	-2.51789
Ross128b	1.09410	0.48964	GJ649c	1.98403	-4.33710
K2-3d	1.14271	0.22599	EPIC211822797b	1.44305	-1.40301

3.1   Results of Thermal Suitability Score Function

For the sake of clarity, two sets of TSS values (shown in Table 1), one for potentially habitable exoplanets, and the other for non-habitable exoplanets.

The potentially habitable sample in this table mostly consists of exoplanets that have been gaining a lot of popularity as potentially habitable worlds. The non-habitable samples are mostly chosen at random, except the TRAPPIST-1 planets, which we included for the sake of completeness with respect to the TRAPPIST-1 potentially habitable samples. Consider the planets TRAPPIST-1 d and b. The differences in sign indicate clear demarcation between the two different classes of habitability. In stark contrast, S.Temp based classification shall place TRAPPIST-1 d and b in the same class because of the proximity of the decision boundary (both TRAPPIST-1 d and b having the same sign as well as close in magnitude). The TSS, unlike S.Temp can thus bolster the discrimination capability (change in sign generates a non-ambiguous separating hyperplane) of a habitability classifier. The variation of the scores from TRAPPIST-1 b to h are reflective of the knowledge gained from ongoing research on the TRAPPIST-1 planets [6, 7].

4   Conclusion and Future Work

This is a metric which is developed using ML and appropriate feature extraction. The method takes our current knowledge and uses it to discriminate and gauge the potential of incoming samples. Although optimization-based approaches have been proposed by [1] and [8], an optimization of an error function in a habitability metric has not been explored before. As it is inherently based on ML, we can increase the number of parameters as long as the notion of linear separability is maintained.

The value of this metric can only be less than 1 for all planets whose surface temperature are not equal to 1 (in EU). The consequence of this is that the value of TSS for only the Earth is equal to 1, and at this point in time, every other planet (which is a part of the PHL-EC) has a TSS of less than one. In addition to that, the *hard-boundary* aspect of SVMs is used to provide results which are negative for non-habitable planets. Conclusively, the negative sign is an out-of-the-box indicator that a planet may not be thermally suitable for habitability. Samples close to the hyperplane may be ambiguous or erroneous; in this model, the hyperplane itself does not perfectly divide the dataset into perfect class-wise partitions, but provides a best-case discriminator. Some of the salient features of TSS are:

1. *Unidirectional Similarity Values*: The value of this metric can only be less than 1 for all planets whose S. Temp values are different from Earth. It doesn't matter if the value is greater or lesser: if it is different, then the value is below that of Earth.
2. *Positive and Negative Values*: Notice that in Table 1, in most places, the sign of the TSS has matched what we know about the habitability potential of these planets. Negative represents non-habitable, and from what we know of the planets in the TRAPPIST-1 system with negative values of TSS, they're not potentially habitable. Additionally, points on the hyperplane will have a

TSS value of zero, and as there are no planets which themselves lie on the maximum margin hyperplane, no planet may have a zero value.

3. *Learning from Example*: This is a metric which is developed *using* ML and appropriate feature extraction. The method takes our current knowledge and uses it to discriminate and gauge the potential of incoming samples.

4. *Tackling Skewness*: From Figs. 1 and 2, we see that the distribution of the habitable samples in the feature space is not symmetric, but there exists a skewness. As a consequence of this, the separating hyperplane is not parallel to the x-axis. However, by thus using the separating hyperplane as a reference boundary, we can equitably judge the samples notwithstanding their respective surface temperatures being lesser than or greater than that of Earth.

5. *Scalable*: As it is inherently based on ML, we can increase the number of parameters as long as the notion of linear separability is maintained.

References

1. Bora, K., Saha, S., Agrawal, S., Safonova, M., Routh, S., Narasimhamurthy, A.: CD-HPF: new habitability score via data analytic modeling. Astron. Comput. **17**, 129–143 (2016)

2. Saha, S., et al.: Theoretical validation of potential habitability via analytical and boosted tree methods: an optimistic study on recently discovered exoplanets. Astron. Comput. **23**, 141–150 (2018)

3. Basak, S., et al.: CEESA meets machine learning: a constant elasticity earth similarity approach to habitability and classification of exoplanets. Astron. Comput. **30**, 100335 (2020)

4. Méndez, A.: The habitable exoplanets catalog (2018)

5. Vapnik, V.N., Chervonenkis, A.Y.: On a class of perceptrons. Autom. Remote Control **1**(25), 103–109 (1964)

6. Barr, A.C., Dobos, V., Kiss, L.L.: Interior structures and tidal heating in the TRAPPIST-1 planets. Astron. Astrophys. **613**, A37 (2017)

7. de Wit, J., et al.: Atmospheric reconnaissance of the habitable-zone earth-sized planets orbiting TRAPPIST-1. Nat. Astron. **2**, 214–219 (2018)

8. Saha, S., et al.: Theoretical Validation of Potential Habitability via Analytical and Boosted Tree Methods: An Optimistic Study on Recently Discovered Exoplanets. arxiv e-prints, December 2017

Author Index

Printed in the United States
By Bookmasters